ASE Test Preparation

Automotive Technician Certification Series

Auto Maintenance and Light Repair (G1)

ASE Test Preparation

Automotive Technician Certification Series

Auto Maintenance and Light Repair (G1)

CENGAGE
Learning·

Australia • Brazil • Mexico • Singapore • United Kingdom • United States

ASE Test Preparation: Automotive Technician Series, Auto Maintenance and Light Repair (G1)

Vice President, Technology and Trades Professional Business Unit: Gregory L. Clayton

Director of Building Trades and Transportation Training: Taryn Zlatin McKenzie

Product Manager: Katie McGuire

Associate Content Developer: Jenn Wheaton

Director of Marketing: Beth A. Lutz

Senior Marketing Manager: Jennifer Barbic

Senior Production Director: Wendy Troeger

Production Manager: Sherondra Thedford

Content Production Management and Composition: PreMediaGlobal

Senior Art Director: Benj Gleeksman

Section Opener Image: © Baloncici/ www.shutterstock.com

Cover Image(s): Iakov Filimonov, 2012. Used under license from Shutterstock.com

Text Image(s): All illustrations within this title are © 2014 Cengage Learning. ALL RIGHTS RESERVED

For product information and technology assistance, contact us at
Cengage Learning Customer & Sales Support, 1-800-354-9706.

For permission to use material from this text or product, submit all requests online at **www.cengage.com/permissions**. Further permissions questions can be e-mailed to **permissionrequest@cengage.com**.

ISBN-13: 978-1-285-75380-5

ISBN-10: 1-285-75380-1

Cengage Learning
200 First Stamford Place, 4th Floor
Stamford, CT 06902
USA

Cengage Learning is a leading provider of customized learning solutions with office locations around the globe, including Singapore, the United Kingdom, Australia, Mexico, Brazil, and Japan. Locate your local office at: **www.cengage.com/global**.

Cengage Learning products are represented in Canada by Nelson Education, Ltd.

For more information on transportation titles available from Cengage Learning, please visit our website at **www.trainingbay.cengage.com**.

Visit our corporate website at **www.cengage.com**.

Notice to the Reader
Publisher does not warrant or guarantee any of the products described herein or perform any independent analysis in connection with any of the product information contained herein. Publisher does not assume, and expressly disclaims, any obligation to obtain and include information other than that provided to it by the manufacturer. The reader is expressly warned to consider and adopt all safety precautions that might be indicated by the activities described herein and to avoid all potential hazards. By following the instructions contained herein, the reader willingly assumes all risks in connection with such instructions. The publisher makes no representations or warranties of any kind, including but not limited to, the warranties of fitness for particular purpose or merchantability, nor are any such representations implied with respect to the material set forth herein, and the publisher takes no responsibility with respect to such material. The publisher shall not be liable for any special, consequential, or exemplary damages resulting, in whole or part, from the readers' use of, or reliance upon, this material.

Printed at CLDPC, USA, 05-17

Table of Contents

SECTION 6 **Answer Keys and Explanations 154**

SECTION 7 **Appendices 283**

Preface

Delmar, a part of Cengage Learning, is very pleased that you have chosen to use our ASE Test Preparation Guide to help prepare yourself for the Auto Maintenance and Light Repair (G1) ASE certification examination. This guide is designed to help prepare you for your actual exam by providing you with an overview and introduction of the testing process, introducing you to the task list for the Auto Maintenance and Light Repair (G1) certification exam, giving you an understanding of what knowledge and skills you are expected to have in order to successfully perform the duties associated with each task area, and providing you with several preparation exams designed to emulate the live exam content in hopes of assessing your overall exam readiness.

If you have a basic working knowledge of the discipline you are testing for, you will find this book is an excellent guide, helping you understand the "must know" items needed to successfully pass the ASE certification exam. This manual is not a textbook. Its objective is to prepare the individual who has the existing requisite experience and knowledge to attempt the challenge of the ASE certification process. This guide cannot replace the hands-on experience and theoretical knowledge required by ASE to master the vehicle repair technology associated with this exam. If you are unable to understand more than a few of the preparation questions and their corresponding explanations in this book, it could be that you require either more shop-floor experience or further study.

This book begins by providing an overview of, and introduction to, the testing process. This section outlines what we recommend you do to prepare, what to expect on the actual test day, and overall methodologies for your success. This section is followed by a detailed overview of the ASE task list to include explanations of the knowledge and skills you must possess to successfully answer questions related to each particular task. After the task list, we provide six sample preparation exams for you to use as a means of evaluating areas of understanding, as well as areas requiring improvement in order to successfully pass the ASE exam. Delmar is the first and only test preparation organization to provide so many unique preparation exams. We enhanced our guides to include this support as a means of providing you with the best preparation product available. Section 6 of this guide includes the answer keys for each preparation exam, along with the answer explanations for each question. Each answer explanation also contains a reference back to the related task or tasks that it assesses. This will provide you with a quick and easy method for referring back to the task list whenever needed. The last section of this book contains blank answer sheet forms you can use as you attempt each preparation exam, along with a glossary of terms.

OUR COMMITMENT TO EXCELLENCE

Thank you for choosing Delmar, Cengage Learning for your ASE test preparation needs. All of the writers, editors, and Delmar staff have worked very hard to make this test preparation guide second to none. We feel confident that you will find this guide easy to use and extremely beneficial as you prepare for your actual ASE exam.

Delmar, Cengage Learning has sought out the best subject-matter experts in the country to help with the development of ASE Test Preparation: Automotive Technician Certification Series, Auto Maintenance and Light Repair, 1st Edition.

Preparation questions are authored and then reviewed by a group of certified, subject-matter experts to ensure the highest level of quality and validity to our product.

If you have any questions concerning this guide or any guide in this series, please visit us on the web at http://www.trainingbay.cengage.com.

For web-based online test preparation for ASE certifications, please visit us on the web at http://www.techniciantestprep.com/ to learn more.

ABOUT THE AUTHOR

Bob Rodriguez is a long-time instructor of automotive electrical, gasoline and diesel injection systems and specializes in alternative fuels and alternative fuel vehicles. Bob's background includes training design and development, training management, delivery, and assessment. Bob is certified by ASE as a Master Automobile Technician (CMAT), and Advanced Level Specialist (L1), a Light Diesel Vehicle Specialist, and Alternate Fuels Technician, and a Parts Specialist (P2). He has been certified as a CNG Cylinder Inspector through CSA, and is trained in advanced automobile electronics and OBD-II systems service.

ABOUT THE SERIES ADVISOR

Randy Nussler became an automotive technician in 1988 and a full time automotive instructor at Midlands Technical College in Columbia, South Carolina in 2005. He earned an Associate in Science Degree in Automotive Technology from New England Institute of Technology. He holds the industry certifications of ASE Master Certified with L1, X1, Undercar Specialist Certified and Subaru Factory Certified Technician.

Randy is active in professional associations as North American Council of Automotive Teachers-South Carolina Chapter (NACAT SC) elected-president and is a member of NACAT, NISOD and IATN. He frequently attends industry training, conferences and serves on a local high schools automotive advisory committee. He is grateful to have been the 2011 Delmar Cengage Jack Erjavec Innovative Instructor Scholarship Award recipient.

The History and Purpose of ASE

ASE began as the National Institute for Automotive Service Excellence (NIASE). It was founded as a non-profit, independent entity in 1972 by a group of industry leaders with the single goal of providing a means for consumers to distinguish between incompetent and competent technicians. It accomplishes this goal through the testing and certification of repair and service professionals. Though it is still known as the National Institute for Automotive Service Excellence, it is now called "ASE" for short.

Today, ASE offers more than 40 certification exams in automotive, medium/heavy duty truck, collision repair and refinish, school bus, transit bus, parts specialist, automobile service consultant, and other industry-related areas. At this time, there are more than 385,000 professionals nationwide with current ASE certifications. These professionals are employed by new car and truck dealerships, independent repair facilities, fleets, service stations, franchised service facilities, and more.

ASE's certification exams are industry-driven and cover practically every on-highway vehicle service segment. The exams are designed to stress the knowledge of job-related skills. Certification consists of passing at least one exam and documenting two years of relevant work experience. To maintain certification, those with ASE credentials must be re-tested every five years.

While ASE certifications are a targeted means of acknowledging the skills and abilities of an individual technician, ASE also has a program designed to provide recognition for highly qualified repair, support, and parts businesses. The Blue Seal of Excellence Recognition Program, allows businesses to showcase their technicians and their commitment to excellence. One of the requirements of becoming Blue Seal recognized is that the facility must have a minimum of 75 percent of their technicians ASE certified. Additional criteria apply, and program details can be found on the ASE website.

ASE recognized that educational programs serving the service and repair industry also needed a way to be recognized as having the faculty, facilities, and equipment to provide a quality education to students wanting to become service professionals. Through the combined efforts of ASE, industry, and education leaders, the non-profit organization entitled the National Automotive Technicians Education Foundation (NATEF) was created in 1983 to evaluate and recognize academic programs. Today more than 2,000 educational programs are NATEF certified.

For additional information about ASE, NATEF, or any of their programs, the following contact information can be used:

National Institute for Automotive Service Excellence (ASE)

101 Blue Seal Drive S.E.

Suite 101

Leesburg, VA 20175

Telephone: 703-669-6600

Fax: 703-669-6123

Website: **www.ase.com**

Overview and Introduction

Participating in the National Institute for Automotive Service Excellence (ASE) voluntary certification program provides you with the opportunity to demonstrate you are a qualified and skilled professional technician who has the "know-how" required to successfully work on today's modern vehicles.

EXAM ADMINISTRATION

> *Note:* After November 2011, ASE will no longer offer paper and pencil certification exams. There will be no Winter testing window in 2012, and ASE will offer and support CBT testing exclusively starting in April 2012.

ASE provides computer-based testing (CBT) exams, which are administered at test centers across the nation. It is recommended that you go to the ASE website at http://www.ase.com and review the conditions and requirements for this type of exam. There is also an exam demonstration page that allows you to personally experience how this type of exam operates before you register.

CBT exams are available four times annually, for two-month windows, with a month of no testing in between each testing window:

- January/February—Winter testing window
- April/May—Spring testing window
- July/August—Summer testing window
- October/November—Fall testing window

Please note, testing windows and timing may change. It is recommended you go to the ASE website at *http://www.ase.com* and review the latest testing schedules.

UNDERSTANDING TEST QUESTION BASICS

ASE exam questions are written by service industry experts. Each question on an exam is created during an ASE-hosted "item-writing" workshop. During these workshops, expert service representatives from manufacturers (domestic and import), aftermarket parts and equipment manufacturers, working technicians, and technical educators gather to share ideas and convert them into actual exam questions. Each exam question written by these experts must then survive review by all members of the group. The questions are designed to address the practical application of repair and diagnosis knowledge and skills practiced by technicians in their day-to-day work.

After the item-writing workshop, all questions are pre-tested and quality-checked on a national sample of technicians. Those questions that meet ASE standards of quality and accuracy are included in the scored sections of the exams; the "rejects" are sent back to the drawing board or discarded altogether.

Depending on the topic of the certification exam, you will be asked between 40 and 80 multiple-choice questions. You can determine the approximate number of questions you can expect to be asked during the Auto Maintenance and Light Repair (G1) certification exam by reviewing the task list in Section 4 of this book. The five-year recertification exam will cover this same content; however, the number of questions for each content area of the recertification exam will be reduced by approximately one-half.

> **Note:** Exams may contain questions that are included for statistical research purposes only. Your answers to these questions will not affect your score, but since you do not know which ones they are, you should answer all questions in the exam.

Using multiple criteria, including cross-sections by age, race, and other background information, ASE is able to guarantee that exam questions do not include bias for or against any particular group. A question that shows bias toward any particular group is discarded.

TEST-TAKING STRATEGIES

Before beginning your exam, quickly look over the exam to determine the total number of questions that you will need to answer. Having this knowledge will help you manage your time throughout the exam to ensure you have enough available to answer all of the questions presented. Read through each question completely before marking your answer. Answer the questions in the order they appear on the exam. Leave the questions blank that you are not sure of and move on to the next question. You can return to those unanswered questions after you have finished the others. These questions may actually be easier to answer at a later time once your mind has had additional time to consider them on a subconscious level. In addition, you might find information in other questions that will help you recall the answers to some of them.

Multiple-choice exams are sometimes challenging because there are often several choices that may seem possible, or partially correct, and therefore it may be difficult to decide on the most appropriate answer choice. The best strategy, in this case, is to first determine the correct answer before looking at the answer options. If you see the answer you decided on, you should still be careful to examine the other answer options to make sure that none seem more correct than yours. If you do not know or are not sure of the answer, read each option very carefully and try to eliminate those options that you know are incorrect. That way, you can often arrive at the correct choice through a process of elimination.

If you have gone through the entire exam, and you still do not know the answer to some of the questions, *then guess*. Yes, guess. You then have at least a 25 percent chance of being correct. While your score is based on the number of questions answered correctly, any question left blank, or unanswered, is automatically scored as incorrect.

There is a lot of "folk" wisdom on the subject of test taking that you may hear about as you prepare for your ASE exam. For example, there are those who would advise you to avoid response options that use certain words such as *all, none, always, never, must,* and *only,* to name a few. This, they claim, is because nothing in life is exclusive. They would advise you to choose response options that use words that allow for some exception, such as *sometimes, frequently, rarely, often, usually, seldom,* and *normally.* They would also advise you to avoid the first and last option (A or D) because exam writers, they feel, are more comfortable if they put the correct answer in the middle (B or C) of the choices. Another recommendation often offered is to select the option that is either shorter or longer than the other three choices because it is more likely to be correct. Some would advise you to never change an answer since your first intuition is usually correct. Another area of "folk" wisdom focuses specifically on any repetitive patterns created by your question responses (e.g., A, B, C, A, B, C, A, B, C).

Many individuals may say that there are actual grains of truth in this "folk" wisdom, and whereas with some exams, this may prove true, it is not relevant in regard to the ASE certification exams. ASE validates all exam questions and test forms through a national sample of technicians, and only those questions and test forms that meet ASE standards of quality and accuracy are included in the scored sections of the exams. Any biased questions or patterns are discarded altogether, and therefore, it is highly unlikely you will experience any of this "folk" wisdom on an actual ASE exam.

PREPARING FOR THE EXAM

Delmar, Cengage Learning wants to make sure we are providing you with the most thorough preparation guide possible. To demonstrate this, we have included hundreds of preparation questions in this guide. These questions are designed to provide as many opportunities as possible to prepare you to successfully pass your ASE exam. The preparation approach we recommend and outline in this book is designed to help you build confidence in demonstrating what task area content you already know well while also outlining what areas you should review in more detail prior to the actual exam.

We recommend that your first step in the preparation process should be to thoroughly review Section 3 of this book. This section contains a description and explanation of the type of questions you'll find on an ASE exam.

Once you understand how the questions will be presented, we then recommend that you thoroughly review Section 4 of this book. This section contains information that will help you establish an understanding of what the exam will be evaluating, and specifically, how many questions to expect in each specific task area.

As your third preparatory step, we recommend you complete your first preparation exam, located in Section 5 of this book. Answer one question at a time. After you answer each question, review the answer and question explanation information located in Section 6. This section will provide you with instant response feedback, allowing you to gauge your progress, one question at a time, throughout this first preparation exam. If after reading the question explanation you do not feel you understand the reasoning for the correct answer, go back and review the task list overview (Section 4) for the task that is related to that question. Included with each question explanation is a clear identifier of the task area that is being assessed (e.g., Task A.1). If at that point you still do not feel you have a solid understanding of the material, identify a good source of information on the topic, such as an educational course, textbook, or other related source of topical learning, and do some additional studying.

After you have completed your first preparation exam and have reviewed your answers, you are ready to complete your next preparation exam. A total of six practice exams are available in Section 5 of this book. For your second preparation exam, we recommend that you answer the questions as if you were taking the actual exam. Do not use any reference material or allow any interruptions in order to get a feel for how you will do on the actual exam. Once you have answered all of the questions, grade your results using the Answer Key in Section 6. For every question that you gave an incorrect answer to, study the explanations to the answers and/or the overview of the related task areas. Try to determine the root cause for missing the question. The easiest thing to correct is learning the correct technical content. The hardest things to correct are behaviors that lead you to an incorrect conclusion. If you knew the information but still got the question incorrect, there is likely a test-taking behavior that will need to be corrected. An example of this would be reading too quickly and skipping over words that affect your reasoning. If you can identify what you did that caused you to answer the question incorrectly, you can eliminate that cause and improve your score.

Here are some basic guidelines to follow while preparing for the exam:

- Focus your studies on those areas you are weak in.
- Be honest with yourself when determining if you understand something.
- Study often but for short periods of time.
- Remove yourself from all distractions when studying.
- Keep in mind that the goal of studying is not just to pass the exam; the real goal is to learn.
- Prepare physically by getting a good night's rest before the exam, and eat meals that provide energy but do not cause discomfort.
- Arrive early to the exam site to avoid long waits as test candidates check in.
- Use all of the time available for your exams. If you finish early, spend the remaining time reviewing your answers.
- Do not leave any questions unanswered. If absolutely necessary, guess. All unanswered questions are automatically scored as incorrect.

Here are some items you will need to bring with you to the exam site:

- A valid government or school-issued photo ID
- Your test center admissions ticket
- A watch (not all test sites have clocks)

Note: Books, calculators, and other reference materials are not allowed in the exam room. The exceptions to this list are English-Foreign dictionaries or glossaries. All items will be inspected before and after testing.

WHAT TO EXPECT DURING THE EXAM

When taking a CBT exam, as soon as you are seated in the testing center, you will be given a brief tutorial to acquaint you with the computer-delivered test prior to taking your certification exam(s). The CBT exams allow you to select only one answer per question. You can also change your answers as many times as you like. When you select a second answer choice, the CBT will automatically unselect your first answer choice. If you want to skip a question to return to later, you can utilize the "flag" feature, which will allow you to quickly identify and review questions whenever you are ready. Prior to completing your exam, you will also be provided with an opportunity to review your answers and address any unanswered questions.

TESTING TIME

Each individual ASE CBT exam has a fixed time limit. Individual exam time will vary based upon exam area, and will range anywhere from a half hour to two hours. You will also be given an additional 30 minutes beyond what is allotted to complete your exams to ensure you have adequate time to perform all necessary check-in procedures, complete a brief CBT tutorial, and potentially complete a post-test survey.

You can register for and take multiple CBT exams during one testing appointment. The maximum time allotment for a CBT appointment is four and a half hours. If you happen to register for so many exams that you will require more time than this, your exams will be scheduled into multiple appointments. This could mean that you have testing on both the morning and afternoon of the same day, or they could be scheduled on different days, depending on your personal preference and the test center's schedule.

It is important to understand that if you arrive late for your CBT test appointment, you will not be able to make up any missed time. You will only have the scheduled amount of time remaining in your appointment to complete your exam(s).

Also, while most people finish their CBT exams within the time allowed, others might feel rushed or not be able to finish the test, due to the implied stress of a specific, individual time limit allotment. Before you register for the CBT exams, you should review the number of exam questions that will be asked along with the amount of time allotted for that exam to determine whether you feel comfortable with the designated time limitation or not.

As an overall time management recommendation, you should monitor your progress and set a time limit you will follow with regard to how much time you will spend on each individual exam question. This should be based on the total number of questions you will be answering.

Also, it is very important to note that if for any reason you wish to leave the testing room during an exam, you must first ask permission. If you happen to finish your exam(s) early and wish to leave the testing site before your designated session appointment is completed, you are permitted to do so only during specified dismissal periods.

UNDERSTANDING HOW YOUR EXAM IS SCORED

You can gain a better perspective about the ASE certification exams if you understand how they are scored. ASE exams are scored by an independent organization having no vested interest in ASE or in the automotive industry. With CBT exams, you will receive your exam scores immediately.

Each question carries the same weight as any other question. For example, if there are 50 questions, each is worth 2 percent of the total score.

Your exam results can tell you:

- Where your knowledge equals or exceeds that needed for competent performance, or
- Where you might need more preparation.

Your ASE exam score report is divided into content "task" areas; it will show the number of questions in each content area and how many of your answers were correct. These numbers provide information about your performance in each area of the exam. However, because there may be a different number of questions in each content area of the exam, a high percentage of correct answers in an area with few questions may not offset a low percentage in an area with many questions.

It should be noted that one does not "fail" an ASE exam. The technician who does not pass is simply told "More Preparation Needed." Though large differences in percentages may indicate problem areas, it is important to consider how many questions were asked in each area. Since each exam evaluates all phases of the work involved in a service specialty, you should be prepared in each area. A low score in one area could keep you from passing an entire exam. If you do not pass the exam, you may take it again at any time it is scheduled to be administered.

There is no such thing as average. You cannot determine your overall exam score by adding the percentages given for each task area and dividing by the number of areas. It doesn't work that way because there generally are not the same number of questions in each task area. A task area with 20 questions, for example, counts more toward your total score than a task area with 10 questions.

Your exam report should give you a good picture of your results and a better understanding of your strengths and areas needing improvement for each task area.

Types of Questions on an ASE Exam

Understanding not only what content areas will be assessed during your exam, but how you can expect exam questions to be presented will enable you to gain the confidence you need to successfully pass an ASE certification exam. The following examples will help you recognize the types of question styles used in ASE exams and assist you in avoiding common errors when answering them.

Most initial certification tests are made up of between 40 and 80 multiple-choice questions. The five-year recertification exams will cover the same content as the initial exam; however, the actual number of questions for each content area will be reduced by approximately one-half. Refer to Section 4 of this book for specific details regarding the number of questions to expect during the initial Engine Repair (A1) certification exam.

Multiple-choice questions are an efficient way to test knowledge. To correctly answer them, you must consider each answer choice as a possibility, and then choose the answer choice that *best* addresses the question. To do this, read each word of the question carefully. Do not assume you know what the question is asking until you have finished reading the entire question.

About 10 percent of the questions on an actual ASE exam will reference an illustration. These drawings contain the information needed to correctly answer the question. The illustration should be studied carefully before attempting to answer the question. When the illustration is showing a system in detail, look over the system and try to figure out how the system works before you look at the question and the possible answers. This approach will ensure that you do not answer the question based upon false assumptions or partial data, but instead have reviewed the entire scenario being presented.

MULTIPLE-CHOICE/DIRECT QUESTIONS

The most common type of question used on an ASE exam is the direct multiple-choice style question. This type of question contains an introductory statement, called a stem, followed by four options: three incorrect answers, called distracters, and one correct answer, the key.

When the questions are written, the point is to make the distracters plausible to draw an inexperienced technician to inadvertently select one of them. This type of question gives a clear indication of the technician's knowledge.

Here is an example of a direct style question:

1. Which of the following would be used to check for rotor runout on a disc brake system?

    A. Dial indicator
    B. Vernier caliper
    C. Outside micrometer
    D. Inside micrometer

TASK E.22

Answer A is correct. A dial indicator is used to measure rotor runout.

Answer B is incorrect. A Vernier caliper is not the correct instrument to use for measuring runout of a rotor.

Answer C is incorrect. An outside micrometer may be used to measure rotor thickness, but not runout.

Answer D is incorrect. An inside mic is not used to determine rotor runout.

COMPLETION QUESTIONS

A completion question is similar to the direct question except the statement may be completed by any one of the four options to form a complete sentence.

Here is an example of a completion question:

TASK A.14

2. Throttle body cleaning should be done by using:

    A.   soap and water.

    B.   silicone spray

    C.   WD-40®

    D.   OEM approved cleaner.

Answer A is incorrect. Soap and water will not be effective for removing carbon and could harm the sensitive MAF sensor if so equipped.

Answer B is incorrect. Silicone spray is a lubricant, not a cleaner.

Answer C is incorrect. WD-40 is also a lubricant, not a cleaner.

Answer D is correct. Only OEM approved cleansers should be used to clean a throttle body.

TECHNICIAN A, TECHNICIAN B QUESTIONS

This type of question is usually associated with an ASE exam. It is, in fact, two true-false statements grouped together, such as: "Technician A says…" and "Technician B says…", followed by "Who is correct?"

In this type of question, you must determine whether either, both, or neither of the statements are correct. To answer this type of question correctly, you must carefully read each technician's statement and judge it on its own merit.

Sometimes this type of question begins with a statement about some analysis or repair procedure. This statement provides the setup or background information required to understand the conditions about which Technician A and Technician B are talking, followed by two statements about the cause of the concern, proper inspection, identification, or repair choices.

Analyzing this type of question is a little easier than the other types because there are only two ideas to consider, although there are still four choices for an answer.

Again, Technician A, Technician B questions are really double true-or-false questions. The best way to analyze this type of question is to consider each technician's statement separately. Ask yourself, "Is A true or false? Is B true or false?" Once you have completed this individual evaluation of each answer choice, you will have successfully determined the correct answer choice for the question, "Who is correct?"

An important point to remember is that an ASE Technician A, Technician B question will never have Technician A and B directly disagreeing with each other. That is why you must evaluate each statement independently.

An example of a Technician A/Technician B style question looks like this:

1. A water pump is being replaced on an engine. Technician A says that on some engines a serpentine belt must first be removed. Technician B says that on some engines the timing belt must first be removed. Who is correct?

    A.   A only
    B.   B only
    C.   Both A and B
    D.   Neither A nor B

Answer A is incorrect. Technician B is also correct.

Answer B is incorrect. Technician A is also correct.

TASK A.11

Answer C is correct. Both Technicians are correct. Some engines have a serpentine belt driven water pump. On other engines the water pump may be driven by a timing belt.

Answer D is incorrect. Both Technicians A and B are making correct statements.

EXCEPT QUESTIONS

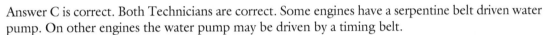

Another type of question type used on the ASE exams contains answer choices that are all correct except for one. To help easily identify this type of question, whenever it is presented in an exam, the word "EXCEPT" will always be displayed in capital letters. Furthermore, a cautionary statement will alert you to the fact that the next question is different from the ones otherwise found in the exam. With the EXCEPT type of question, only one *incorrect* choice will actually be listed among the options, and that incorrect choice will be the key to the question. That is, the incorrect statement is counted as the correct answer for that question.

Be careful to read these question types slowly and thoroughly; otherwise, you may overlook what the question is actually asking and answer the question by selecting the first correct statement.

An example of this type of question would appear as follows:

1. A diesel engine is being checked for an engine oil leak. Any of these could be used to help locate the source of the leak EXCEPT:

    A.   Black light.
    B.   White powder.
    C.   Vacuum gauge.
    D.   Oil dye.

Answer A is incorrect. A black light can be used to help locate the source of a leak.

Answer B is incorrect. White powder can be used to help locate the source of a leak.

Answer C is correct. A vacuum gauge is not used to locate engine oil leaks.

Answer D is incorrect. Oil dye may be used to help locate an engine oil leak.

TASK A.2

LEAST LIKELY QUESTIONS

LEAST LIKELY questions are similar to EXCEPT questions. Look for the answer choice that would be the LEAST LIKELY cause (most incorrect) of the described situation. To help easily identify these types of question, whenever they are presented in an exam the words "LEAST LIKELY" will always be displayed in capital letters. In addition, you will be alerted before a LEAST LIKELY question is posed. Read the entire question carefully before choosing your answer.

An example of this type of question is shown here:

1. A vehicle equipped with a diesel engine overheats when pulling a trailer. Which of these would be the LEAST LIKELY cause?

    A. Slipping fan clutch

    B. Seized fan clutch

    C. Restricted charge air cooler

    D. Restricted radiator

A. 7
TASK A.12

Answer A is incorrect. A slipping fan clutch may not fully engage and would fail to provide sufficient air flow across the radiator to keep the engine cool.

Answer B is correct. A seized fan clutch would run all the time; this may cause a low power complaint, but **would not** cause the engine to overheat.

Answer C is incorrect. A restricted charge air cooler would also restrict the air flow across the radiator. This could result in an engine overheating condition.

Answer D is incorrect. A restricted radiator could result in an overheated engine.

SUMMARY

The question styles outlined in this section are the only ones you will encounter on any ASE certification exam. ASE does not use any other types of question styles, such as fill-in-the-blank, true/false, word-matching, or essay. ASE also will not require you to draw diagrams or sketches to support any of your answer selections, although any of the described question styles may include illustrations, charts, or schematics to clarify a question. If a formula or chart is required to answer a question, it will be provided for you.

Task List Overview

INTRODUCTION

This section of the book outlines the content areas or *task list* for this specific certification exam, along with a written overview of the content covered in the exam.

The task list describes the actual knowledge and skills necessary for a technician to successfully perform the work associated with each skill area. This task list is the fundamental guideline you should use to understand what areas you can expect to be tested on, as well as how each individual area is weighted to include the approximate number of questions you can expect to be given for that area during the ASE certification exam. It is important to note that the number of exam questions for a particular area is to be used as a guideline only. ASE advises that the questions on the exam may not equal the number specifically listed on the task list. The task lists are specifically designed to tell you what ASE expects you to know how to do and to help prepare you to be tested.

Similar to the role this task list will play in regard to the actual ASE exam, Delmar, Cengage Learning has developed six preparation exams, located in Section 5 of this book, using this task list as a guide. It is important to note that although both ASE and Delmar, Cengage Learning use the same task list as a guideline for creating these test questions, none of the test questions you will see in this book will be found in the actual, live ASE exams. This is true for any test preparatory material you use. Real exam questions are *only* visible during the actual ASE exams.

Task List at a Glance

The Auto Maintenance and Light Repair (G1) task list focuses on 7 core areas, and you can expect to be asked a total of approximately 55 questions on your certification exam, broken out as outlined:

- A. Engine Systems (9 Questions)
- B. Automatic Transmission/Transaxle (4 Questions)
- C. Manual Drive Train and Axles (6 Questions)
- D. Suspension and Steering (13 Questions)
- E. Brakes (11 Questions)
- F. Electrical (8 Questions)
- G. Heating, Ventilation, and Air Conditioning (4 Questions)

Based upon this information, below is a general graphical guideline demonstrating which areas will have the most focus on the actual certification exam. This data may help you prioritize your time when preparing for the exam.

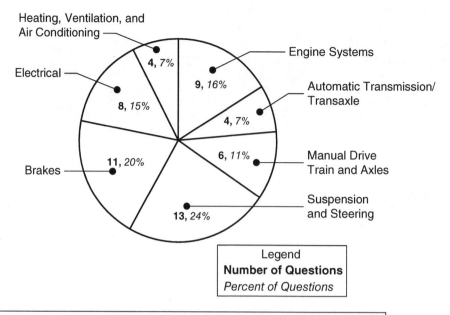

Legend
Number of Questions
Percent of Questions

Note: The actual number of questions you will be given on the ASE certification exam may vary slightly from the information provided in the task list, as exams may contain questions that are included for statistical research purposes only. Do not forget that your answers to these research questions will not affect your score.

TASK LIST AND OVERVIEW

A. Engine Systems (9 questions)

1. Verify driver's complaint and/or road test vehicle; determine necessary action. Utilize service manuals, technical service bulletins (TSBs), and product information.

Before starting to work on any vehicle, it is important to gather as much information as you can in order to avoid wasting time and effort. You will need to read the customer's concern written on the repair order carefully to determine just what needs to be addressed. Perhaps the vehicle is simply in for scheduled maintenance, or perhaps for an unscheduled diagnosis and repair. Sometimes interviewing the customer and going for a test drive are the best ways to investigate the driver's concerns. Perhaps simply understanding and following the instructions in the vehicle owner's manual will solve the problem.

Once the vehicle owner's complaint is verified, consult the latest Technical Service Bulletins (TSBs) for the latest service manual updates and/or information related to the customer's specific complaint. Often consulting on-line forums or services such as iATN® or Identifix® will lead you to an experience-based solution to the customer's vehicle problem.

Finally, consider consulting the TSBs, catalogs, and manuals provided by parts suppliers for additional service tips related to their specific products.

2. Inspect engine assembly for fuel, oil, coolant, and other leaks; determine necessary action.

A routine engine inspection should include looking both underhood and undercar for gasoline or diesel fuel, oil, coolant, or other forms of leakage. Most fluids have distinct odors and colors that help to identify what they are.

Valve covers and front or rear crankshaft seals are notorious places from which high mileage engines may leak oil. Coolant may seep from a leaking head gasket, thermostat housing, or freeze plug. In severe cases, a cracked head or engine block may exhibit a coolant or oil leak. In cases other than leaking gaskets, a major repair requiring head removal or removing the crankshaft may be needed to fix the leakage problem.

3. Check for abnormal engine noises.

Internal gasoline engines make all sorts of noises during normal operation, ranging from combustion noise, to exhaust noise to light or heavy mechanical sounds. These include squealing, rattling, ticking, clunking and anywhere in between. Normal noises may include light tappet noise or the sound of belts running in their pulleys.

When the starter is engaged, the engine should crank at a reasonably sufficient rpm (see manufacturer's specifications). Most manufacturers of electronically controlled gasoline or diesel fuel-injected engines will not inject fuel until about 150 rpm. The cranking speed can be easily verified with an electronic scan tool connected to the engine.

During a test drive, listen for abnormal combustion sounds like pinging or detonation which can be caused by any number of faults, from bad/contaminated gasoline to a faulty EGR valve system, or an intake leak which may be heard as a "whistling" sound.

Engine noises may originate from many areas. Engine crankshaft bearing noises are usually located lower in the engine and are referred to as a knock, occurring at crankshaft speed. The sources of these noises can be determined by raising engine rpm and noting any change in noise intensity. They are also diagnosed by removing combustion from the cylinder. In earlier engines this was done by pulling a spark plug wire from the spark plug, or loosening a high-pressure diesel injection line. In electronically controlled engines this is done by using the electronic scan tool in an Interactive Diagnosis mode. If the noise goes away when the spark or fuel delivery is removed, then the noise is usually considered to be a rod bearing problem. A double knock that does not go away when the fuel is removed from the combustion chamber is usually a piston pin (wrist pin).

Noises that come from the valve train are generally located higher in the engine and occur at ½ crankshaft speed. These noises are usually lighter in sound and generally referred to as a ticking or clicking. Causes of valve train noises may include worn cam lobes, lifters (followers), rocker arms, and bent push rods. Loud ticking noises or rattling could also mean low oil pressure being supplied to the valve train, or collapsed lifters due to incorrect motor oil or lack of maintenance. Squeals may mean a dry or worn serpentine or "V" belt. Listen for a noisy belt tensioner or pulleys requiring replacement.

All of these kinds of noises may best be located and isolated by using a stethoscope, or simply by probing with a piece of rubber tubing placed to one's ear. Use caution when doing so to avoid tangling with moving engine parts.

4. Inspect and replace pans and covers.

Check the oil pan and valve cover gaskets for leaks. Sometimes loosening and retorquing their bolts in the correct sequence will temporarily fix a leak, but ultimately a dry or cracked gasket will need to be replaced.

Modern vehicles use plastic instead of metal for valve covers and pans, so be extra careful when working around these parts. If the valve covers or pans are cracked, do not attempt to repair them; replace them with the correct replacement part(s).

5. Change engine oil and filter; reset oil life monitor.

Modern car engines generally require oil changes at 5,000 mile intervals and higher. Follow the schedule in the owner's manual, which is based on the type of driving done. When changing the motor oil on a vehicle, try to do it when the engine is warm. This ensures that the used oil will drain more quickly and thoroughly.

Remove the oil fill cap and the oil drain plug to drain the used motor oil into a drain pan. While the old oil is draining, remove the used oil filter with the appropriate filter socket or strap wrench. Be sure the old filter gasket is removed and clean the filter gasket surface with a lint free cloth. Add fresh oil to the new filter and apply a light coating of motor oil to the gasket before installing it. Tighten the oil filter firmly by hand. Using a new gasket, replace the oil pan drain plug. Put the recommended type and amount of oil into the engine as per the OEM maintenance schedule. If applicable, reset the oil change reminder in the driver's information center display.

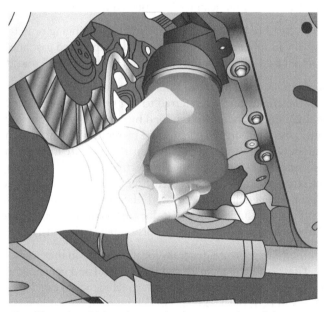

The filter should be changed whenever the oil is changed. Make sure you drain and recycle used oil filters in an 'environmentally friendly' manner.

6. Inspect and test radiator, heater core, pressure cap, and coolant recovery system; determine needed repairs; perform cooling system pressure and dye tests.

It is important to periodically inspect and test the cooling system for leaks and for pressure holding capability. When the cooling system is being flushed and the coolant is being replaced, it is a good time for such service. Use a radiator/radiator cap tester with the proper adaptor to pressurize the cooling system (only when cool) and determine if any leaks exist. Test the cap to ensure the cap is holding, and relieving, pressure properly. Do not over-pressurize the system, but do look for signs of falling pressure, which would indicate a leak.

Check for telltale residue left around the upper and lower chambers of the radiator and elsewhere. Check for a loss of (typically green or orange) coolant at the evaporator drain hose under the vehicle, which would indicate a leaking heater core. Some coolant recovery bottles remain pressurized during hot engine operation, so check the plastic recovery bottles for leaks as well.

If a coolant leak is elusive, ultraviolet dye can be added to the cooling system, after which the system is inspected using a UV lamp.

> **Note:** Coolant may also leak into the intake or exhaust system if a cylinder head or engine block is cracked. In these cases, an abnormal amount of white smoke may be seen coming from the exhaust pipe.

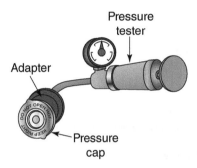

7. Inspect, replace, and adjust drive belt(s), tensioner(s), and pulleys.

There are numerous "V" type and serpentine belts used under the hood of today's vehicles. Such belts do not last forever. During a vehicle inspection, check them for cracking, splitting, dryness, and wear. Serpentine belt manufacturers tell us that modern belt composition does not show the usual signs of dryness and cracking as belts age. The proper way to test serpentine belts is by using a wear gauge (see image).

When replacing a belt, make sure it is properly routed by following the diagram found on the radiator shroud or the underside of the hood. Also be certain the grooves of a serpentine belt are properly set in the pulleys.

Proper belt tension may be established automatically by a belt tensioner, or in some cases, the tension must be set by the technician using a belt tension gauge or another method such as checking deflection with a ruler.

A noisy or squealing belt may need adjustment or replacement. A tensioner or pulley that is noisy may be a part that is waiting to fail. Replace such parts as soon as possible to avoid an inconvenient breakdown. Some tensioners are especially designed to take up and release tension as the belt rotates, so make sure the proper replacement part number is ordered and installed by comparing the old part with the replacement part.

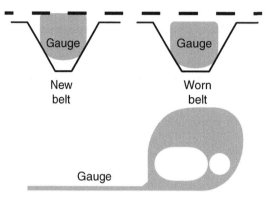

New belt

Worn belt

Gauge

Today's serpentine belts do not always look worn, but still need to be periodically checked. Use a serpentine belt wear gauge like this to determine if the belt is still good.

8. Inspect and replace engine cooling system and heater system hoses, pipes, and fittings.

Engine cooling system and heater hoses do not last forever. Some maintenance schedules call for periodic replacement of hoses even if the parts do not appear worn or ready to fail. Never attempt to replace hoses when the engine is hot! Open the radiator drain cock or loosen a lower radiator hose to drain the system enough to remove and replace the needed hose. Be sure to catch all used coolant in a pan and recycle it.

Route heater hoses the way the OEM intended. They should not be too long or too short. Heater hoses must not be allowed to rest on hot engine components. Some heater systems use metal pipes or specially shaped fittings. Be sure to replace these with OEM or equivalent parts rather than with makeshift or leftover parts.

9. Remove and replace engine thermostat and coolant bypass.

Sometimes a thermostat will fail and not allow the engine to warm up (they usually fail open). In some cases a failed thermostat will cause the MIL to illuminate because some OBD II monitors will only run when the engine is fully warmed up. Only replace a thermostat with an OEM or equivalent part. When installing it, be certain it is positioned properly and seated correctly with the weep hole or jiggle valve upwards. Use a new gasket held in place with a THIN coating of sealant to hold it in place while installing the housing bolts.

Bypass hose replacement may be performed. When replacing them, use OEM or equivalent replacements wherever possible to avoid problems with kinking for deforming.

10. Inspect and test coolant; drain, flush, and refill cooling system with recommended coolant; bleed air as required.

Inspect the coolant for rust, scale, corrosion, and other contaminants such as engine oil or automatic transmission fluid. If the coolant is contaminated, the cooling system should be drained and flushed. If oil is floating on top of the coolant, the engine block or head may be cracked, or the automatic transmission cooler may be leaking. (This condition also means there is coolant in the transmission!) If the vehicle has an external engine oil cooler, it may also be a source of oil contamination in the coolant.

Engine coolant consists of antifreeze and water, typically in a 50/50 mix. The antifreeze content may be tested with a hydrometer, to measure the coolant's specific gravity which indicates the ratio of water to antifreeze. The freezing point of the coolant is indicated on the hydrometer float. The OEM-specified antifreeze content must be maintained in the cooling system.

There are various chemical compositions of antifreeze specified by the OEMs for their vehicles, and they should not be mixed or used in incorrect vehicles. The color of the antifreeze (green, yellow/orange, etc.) helps to identify the type of antifreeze. Check your specifications before adding antifreeze to the vehicle cooling system.

Some vehicles may need to have their cooling system vented of air once the system is filled. This may be as simple as opening a vent screw or slightly raising one side of the vehicle with a floor jack in order to purge air from the system. Follow specific OEM procedures to eliminate air pockets.

11. Inspect and replace accessory belt driven water pumps.

Some vehicles use electric water pumps, but most vehicles still use belt-driven water pumps. These pumps will eventually fail with high mileage due to leaking seals or worn bearings. Check for a noisy pump or for coolant dripping from a belt-driven pump, which indicates the pump's bearing or seal is failing.

Older vehicles use a V-belt to drive the water pump from the crankshaft pulley. Most contemporary vehicles drive the water pump either by a crankshaft-driven serpentine belt or toothed timing belt. Carefully inspect the drive belt for excessive tightness or looseness, wear or deterioration. Something as simple as a failed water pump belt or broken belt will quickly disable a vehicle and possibly contribute to serious internal engine damage.

> *Note:* Since in some vehicles removal of the timing belt must be done to get to the engine-driven water pump, the water pump is often changed routinely during a timing belt replacement to save repeated work later should the old pump fail.

12. Confirm fan operation (both electrical and mechanical); inspect fan clutch, fan shroud, and air dams.

The engine crankshaft-driven cooling fan on older vehicles uses a thermostatic clutch to help reduce noise and parasitic horsepower loss when the fan is not needed. Check that the clutch is firm and not dripping oil. The fan also relies on shrouding to help direct ram air through the radiator and through the fan blades even when the vehicle is stationery. Check that the shrouding has not been damaged or removed. Check that the air dam in front of the vehicle has not been damaged or torn off from careless parking habits (against curb stops) or from driving off road.

The same applies to electrically driven and thermostatically controlled fans. Make sure they are properly shrouded and check that the fan operates when the air conditioning is turned on or when the engine is idling and fully warmed up.

13. Verify operation of engine-related warning indicators.

There are various dash instruments and warning lights on the instrument cluster that alert the driver of undesirable or unsafe conditions. Many of these indicators are self-tested whenever the vehicle is started, but the vehicle inspection should include making sure all of these indicators are working properly. For vehicles with instruments for the alternator, water temperature, and oil pressure, make certain they read normally. If "idiot lights" are used in place of the voltmeter/ammeter, oil pressure, and coolant temperature gauges, make certain they function as they should. Check the vehicle owner's manual to determine which lights and gauges are included on the instrument panel cluster for indicating abnormal engine conditions.

Verify that all warning lights and indicators work as they should when the ignition is first turned to ON.

14. Perform air induction/throttle body service.

After many miles, oil and dirt can accumulate in the induction system despite a properly maintained intake system air filter. Accumulations of carbon can build up on the throttle body and even cause the throttle plate to stick. The PCM actuated idle-air control motor (IAC) can become clogged with carbon and sticky residue from the PCV system and cause an upset or surging of the engine's idle speed. Remove the air cleaner and inspect for impurities in the intake; check for a sticking throttle plate as it rests on the throat of the throttle body.

Use only OEM-approved cleaning methods, especially when spray cleaning the intake with solvents. The wrong chemical could ruin the mass airflow sensor (MAF) or other sensitive sensors.

Make certain the throttle and cruise control cables operate smoothly, or that the wiring for the drive-by-wire circuitry is not damaged or faulty in any way. Rodents have been known to nest under the warm hood and feast on such wiring harnesses.

15. Inspect, service, or replace air filter(s), filter housing(s), and air intake system components.

Any routine inspection and maintenance program includes the periodic replacement of the air filter. While the filter is being inspected or replaced, inspect the air box for debris such as leaves and insects. Inspect the wiring used for intake air or battery temperature sensors, and make certain the intake duct clamps are snug and the tubing is not cracked, allowing "false air" or unfiltered air to be admitted into the engine. Inspect and clean or replace the PCV filter in the air box as well.

16. Inspect and replace crankcase ventilation system components.

Make certain the filter in the air filter box is clean. Make certain the PCV tubing leading from the PCV valve to the throttle body is not cracked or broken. If the PCV tubing is cracked or broken, it would admit unfiltered air into the intake system.

Remove and shake the PCV valve to make certain it "rattles." If it does not, clean it with solvent or replace it. If there are oil leaks around the valve cover's PCV tubing, inspect for dry or cracked fittings and replace them as needed.

17. Inspect exhaust system for leaks; check hangers, brackets, and heat shields; determine needed repairs.

Today's vehicle exhaust systems last an amazingly long time because stainless steel components are being used by the OEMs for exhaust system components. Still, they should be inspected at every opportunity. With the engine running at idle, hold a shop rag tightly against the exhaust pipe outlet and listen for any "puffing" or leaks of exhaust from the exhaust plumbing or components.

When the car is on the lift, check for exhaust component damage from road debris or from driving off-road. Tap on the exhaust system with a screwdriver and listen for thin or rusted areas which do not "ring" as metal pipes should. Inspect for loose, torn, or missing exhaust pipe hangers and brackets. Physically shake the exhaust system to make sure components are not striking the suspension, drive train, or other undercar parts.

Check that the catalytic converter is in place and that the heat shield is not missing, damaged, rusted through, or loose. Check for rusting or holes in the muffler, resonator, converter, and piping.

Finally, check that the wiring for the pre- and post- CAT sensors is in good shape.

18. Retrieve and record diagnostic trouble codes (DTCs).

This part of the inspection process is the fun part; that is, when the MLR technician taps into the PCM data stream to see what is really happening. Using the proper scan tool and software for the vehicle under inspection, connect to the DLC and power up the scan tool. Follow the respective scan tool instructions to determine if any history codes are stored that might indicate past repairs or future failures.

If the MIL remains illuminated after the engine is started, retrieve and write down the active/stored codes—DO NOT erase them. Both the active and historic codes will help the diagnostic technicians on staff to determine the root cause(s) for the MIL to be illuminated, and to facilitate the needed repairs.

19. Remove and replace spark plugs; inspect secondary ignition components for wear or damage.

Though spark plugs last far longer than they used to before unleaded gasoline, maintenance schedules still call for replacing the spark plugs at specific intervals—as much as 100,000 miles. At that point, plugs can be difficult to remove, and may require special removal techniques to avoid stripping the threads out of the cylinder heads.

Be sure to check for TSBs before determining to "get them out" (literally at any cost). When replacing plugs, it may be wise to use the OEM brand and type to avoid potential DTCs being set later. Use the proper gapping tool to check/set the spark plug gap to specification, and use the proper wrench to avoid breaking the insulator during the installation. Coil over plugs have made failures due to moisture and dirt less frequent compared to when ignition cables were used, but they too need to be inspected for possible wire fraying or damage.

If the vehicle has distributorless ignition, make sure the primary wiring is secure and that the secondary wires are clean and free of wear. If they appear dried or cracked, replace them as a set with good quality replacements.

If the car is equipped with an ignition distributor, inspect the cap inside and out, and inspect the rotor for corrosion or burn marks. Look for signs of oil inside the distributor, which would indicate a leaking distributor seal, and look for electrical arcing on the inside of the cap, which means it needs replacing.

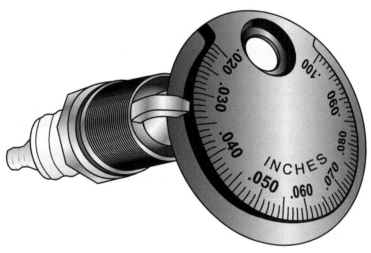

Use a gapping tool to be sure the spark plug gap is set correctly before installing replacement sparkplugs. Some vehicles are sensitive to the brand and type of plugs used, so it may be best to stick with the OEM recommended plugs to avoid potential problems.

20. Inspect fuel tank, filler neck, fuel cap, lines, fittings, and hoses; replace external fuel filter.

If the fuel system is leaking vapors in any way on OBD II vehicles, the MIL will be illuminated for an EVAP system leak. From under the car, inspect the fuel tank for signs of rusting, damage or leaks. Also inspect the filler neck and the cap's seat for damage that would allow vapors to escape to the atmosphere. Inspect the cap to make certain the "O" ring is intact. Check for loose or split EVAP hoses; check for leaks at the external fuel filter and all fuel fittings.

If replacing the external fuel filter, follow the OEM instructions to the letter. A special tool may be required to remove the filter fuel line clamps. Catch any fuel spills in a rag, and make certain the replacement filter is installed facing the correct direction for fuel flow, as is customarily indicted by an arrow pointing towards the engine/outlet side of the filter.

21. Inspect canister, lines/hoses, mechanical and electrical components of the evaporative emissions control system (EVAP).

Inspect all electrical, vacuum, and fuel line connections to EVAP components. As mentioned above, the EVAP system is monitored by the OBD II system. Even minor EVAP leaks will be detected and trip a fault code (a DTC). When this happens, first check that the fuel cap is properly installed on the filler pipe. Second, check for loose or cracked flexible hose/tubing around the fuel canister and other EVAP components both under the hood and under the car. For difficult-to-find EVAP leaks, a "smoke machine" can be used to very slightly pressurize the fuel system with smoke and UV dye. Any leaks will be found where the smoke escapes or leaves a UV sensitive trace of powder.

22. Check and refill diesel exhaust fluid (DEF).

Today's diesel vehicles employ a variety of emission control devices that were unheard of just a few short years ago. One of these is the treatment of NOx through the use of special urea-based fluid injected into the exhaust stream. Selective catalytic reduction (SCR) is used to reduce the amount of NOx released into the air by using the diesel exhaust fluid (DEF) to turn smog-forming NOx into harmless nitrogen and water.

The blue-colored fluid is added periodically to a special container on board the diesel vehicle, and is automatically injected by the computer-controlled emission system. If the fluid runs low, the driver is warned, and if the container becomes seriously low, the engine is de-rated of power to make certain the driver gets the message.

Routine maintenance or inspection of a late model diesel vehicle would certainly include checking and filling as needed the blue-capped DEF container with urea fluid sold by any number of distributors. DEF is sometimes marketed by German OEMs under the name AdBlue®.

B. Automatic Transmission/Transaxle (4 questions)

1. Road test the vehicle to normal operation; retrieve and record diagnostic trouble codes (DTCs).

Unusual automatic transmission noises or smells may be evident and even obvious with the car parked in the shop. Likewise, it is obvious when the car will not move forward in LOW or DRIVE, or backwards in REVERSE. In many cases, however, it would be difficult to verify a customer's concern about the automatic transmission's performance in a vehicle without taking it for a test drive.

While on the road, listen for whining of the pump, be alert for harsh up or downshifting, and determine if shifting occurs at the proper time for given speeds and loads. Perhaps a simple adjustment is all that is needed. On the other hand, it may be that something is broken, causing the transmission to only move forward in LOW, or perhaps the transmission stays in "LIMP" mode.

Let the diagnostic capability of the TCM help determine a troubleshooting strategy. Take careful note of any unusual operation, and once back in the shop, retrieve and save any stored codes in the Transmission Control Module (TCM). To save time and effort, refer to the diagnostic "trouble trees" in the vehicle service manual to determine the proper diagnostic strategy to follow.

2. Determine fluid type, level, and condition.

Pull the dipstick and smell the fluid. Does it smell bad or burnt? How is its color? Is it red as it should normally be, or is it a dirty brown in color? Is the fluid level correct for the temperature at which it is checked? Automatic transmission fluid expands as it gets warm, so do not make the mistake of adding ATF when it is not needed.

Check the level with the vehicle on a level surface and with the A/T warmed up. If the fluid smells burnt, the internal bands and/or clutch surfaces may be burnt, requiring a transmission rebuild or replacement.

Check the transmission fluid level while hot; also check to see if it smells burnt.

3. Inspect transmission for leaks; replace external seals and gaskets.

If the ATF is found to be low, check for leaks at the transmission pan gasket. If a transmission pan gasket is in good condition, some OEMs advocate reusing the gasket more than once. Some OEMs advocate using special sealant rather than a gasket. When using sealant, use it sparingly and make certain it is placed along the outside of the pan's bolt holes to minimize the possibility of sealant getting into the transmission itself.

4. Inspect and replace CV boots, axles, drive shafts, U-joints, drive axle joints, and seals.

Today's front-wheel-drive vehicles often use MacPherson strut design front ends. These rely on constant velocity joints (CV) joints to enable the front wheels to turn sharply and yet drive smoothly without jerkiness or inconsistent torque. Over time, the boots that protect the inner—and especially the outer—CV joints may crack or become torn, allowing water and dirt to contaminate the CV joint inside. Once this happens, it is only a matter of time before the joint fails.

A leaking CV joint boot can be evident due to traces of lubricant splattered on the suspension components nearby. If caught in time, a replacement boot can be installed without disassembling the joint itself, or the driveshaft (half-axle) can be removed from the vehicle and the joint rebuilt. Sometimes it is easier and less expensive to replace the entire axle shaft assembly with a rebuilt unit that includes the inner and outer CV joints.

For rear-wheel-drive (RWD) and four-wheel-drive (4WD) vehicles, driveshafts and drive axles use universal joints for transferring torque at ever-changing angles. These require periodic inspection. Check driveshafts and U-joints for signs of rust, meaning a lack of lubricant. Also check for U-joint needle bearing failure, evident by rust, looseness and play in the joints themselves. When replacing driveshafts, be sure to reinstall them with the same orientation (phasing) as when they came out.

Check transmission/transaxle half-axle seals and output shaft seals for leaks. If they are leaking, check for excessive internal bearing play or for a plugged transmission breather. Replace worn or faulty components as needed.

5. Visually inspect condition of transmission cooling system, lines, and fittings.

A routine undercar inspection includes looking for leaks at the transmission cooler, the cooler lines, and fittings. ATF hoses must allow for engine/transmission movement and may over time become soft and porous. Check for signs of transmission fluid dripping on the floor. Replace components as needed.

6. Inspect and replace power train mounts.

The engine and transmission assembly need to be able to rock with torque reversals associated with accelerating and coasting. This ability is provided by flexible motor and transmission (aka powertrain) mounts. Over time, vehicles tend to drip oil or fluids down onto these mounts, and oil has an especially adverse affect on the flexible rubber used in them.

Check powertrain mounts for swelling or deterioration. Having an engine come loose in a vehicle is obviously not a good thing, and may be evident by the transmission jumping out of gear. Raise the engine or transmission slightly with a jack to determine if the mounts have separated. Replace them if necessary.

7. Replace fluid and filter(s).

Until recently, OEMs have advocated periodic replacement of the automatic transmission filter, especially if the vehicle is used in dusty environments. In many cases, this requires dropping the transmission pan to get to the filter. Since there is likely no drain plug on the A/T, be careful when removing the pan bolts to avoid a major spill onto the shop floor— or on yourself! The pan should only be removed where the environment is free of dirt and dust and the surrounding air is still (no fans blowing). Getting dirt inside the transmission's fine internal moving parts could be worse than living with a dirty transmission filter. While the pan is removed, remove any fine metal particles from the collecting magnet in the pan. Thoroughly clean and dry the pan, and if used, replace the gasket. Tighten the bolts hand tight, then torque them in the proper sequence. Add new ATF as needed to the transmission, making sure it is the proper type. Also be sure to leave some room for fluid expansion once it is warm.

> *Note:* Some contemporary vehicles do not have a means for checking fluid level or replacing a filter. Unless there is a leak, the A/T is deemed to be a maintenance-free component. Always check the OEM service procedures before attempting transmission service.

C. Manual Drive Train and Axles (6 questions)

1. Inspect, adjust, replace, and bleed external hydraulic clutch slave/release cylinder, master cylinder, lines, and hoses; clean and flush hydraulic system; refill with proper fluid.

Many newer vehicles are equipped with a hydraulically actuated clutch (these systems typically use brake fluid). If the clutch does not engage or disengage as it should, check for leaks. Hydraulic clutches with a constant-running release bearing do not require adjustment. Check the system for low fluid or for air in the system. If the fluid level is low, inspect the system for leaks. If fluid is leaking from the master cylinder or slave cylinder, overhaul or replace them as required. If fluid is leaking from the line between the master cylinder and slave cylinder, repair or replace the line.

If air has gotten into the system, check that all fittings are snug. Bleed any air from the hydraulic clutch system, which is similar to bleeding a brake system, by following the vehicle manufacturer's procedures. If the system has been contaminated with the wrong fluid, drain and flush the hydraulic clutch system and fill it with the approved fluid (normally brake fluid) to the correct level.

> *Note:* If the vehicle has a clutch actuated by a cable or linkage, inspect the self-adjusting cable mechanism and /or linkage for wear or damage. If an out-of-adjustment condition exists, adjust the linkage for proper clutch pedal free play.

2. Inspect and replace power train mounts.

As mentioned in Task B.6, the engine and transmission assembly need to be able to rock with torque reversals associated with accelerating and coasting. The ability for this is provided by flexible powertrain mounts. Refer to Task B.6 for inspection and servicing instructions.

3. Inspect, adjust, and replace transmission/transaxle external shifter assembly, shift linkages, brackets, bushings/grommets, pivots, and levers.

The shift linkage or cable adjustment procedure varies depending on whether the transaxle has shift cables or linkages and upon the vehicle make, model, and year. Making the adjustment on a certain vehicle involves removing, reversing, and reinserting the lock pin from the transaxle selector shaft housing. The pin locks the 1–2 shift fork shaft in the neutral position. Next, remove the gearshift knob and console. Then, loosen the selector cable and crossover cable adjusting screws and install a 3/16-inch (4.75 millimeter) drill bit into the adjusting pin openings for each cable. Now, tighten the selector screws on the selector cable and the crossover cable to the specified torque. Finally, remove the adjusting pins, reinstall the lock pin, and install the console and gearshift knob.

Start the engine, fully depress the clutch pedal, and shift the gear selector through all the gear positions while checking for proper shifting without gear clashing. Road test the vehicle and check for proper shifting.

The shift linkage or cable adjustment procedure varies depending on whether the transaxle has shift cables or linkages, and upon the vehicle make, model, and year. Making the linkage adjustment may involve removing, reversing, and reinserting a lock pin from the transaxle selector shaft housing.

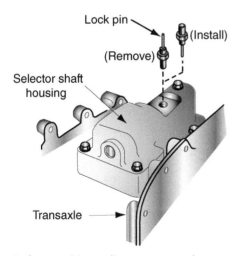

Before making adjustments to the crossover and gear select cables on a manual drivetrain vehicle, some vehicles require that the shift mechanism be locked in neutral. To do this, remove the gearshift knob and console; then remove, reverse and re-install the selector housing lock pin, as shown here.

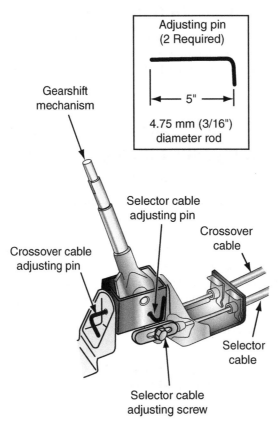

Next, loosen the two cable adjustment screws and insert a **3/16**" (**4.75** mm) "adjusting pin" into each of the two adjusting pin openings, as shown here.

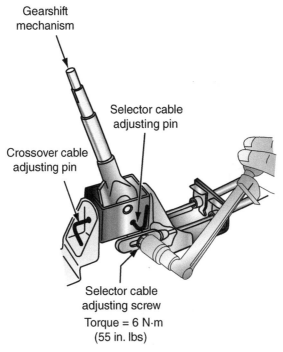

With the adjusting pins still installed, re-tighten and torque the cable adjusting screws to specifications; then remove the adjusting pins, reverse the lock pin to its original position, and re-install the console and gearshift knob.

4. Inspect and replace external seals.

Check the rear main seal of a RWD vehicle for leakage of transmission lubricant. If leaking, check for a worn tailshaft bearing by pushing the drive shaft U-joint up, down, and sideways while observing lateral movement of the tailshaft. If movement is noticed, the rear main bearing may be worn and should be replaced along with a new rear seal. If a FWD transaxle is leaking at the half shafts, use the same procedure to check for bearing wear. Replace the axle bearings and seals if needed.

Check for transmission fluid leaking from the clutch housing. If leaking is detected, remove the inspection plate, if provided, and check for oil from a failed input shaft bearing seal (or a rear main engine oil seal). (Check for an oil-contaminated clutch disc at the same time.)

Also check the transmission cover plate gasket and shift linkage seals for signs of oil leaking. Replace as required. Check for a plugged transmission vent, which could cause internal pressure to force oil past the seals.

5. Check fluid level; refill with fluid.

To check for correct lubricant oil level in a manual transmission, look for a metal screw-in plug, a dipstick, or some other means of gaining access to the transmission lubricant. At least one OEM requires that the speedometer drive gear be removed for checking the manual transmission oil level. Make sure the car is level when checking the fluid level. Remove the side plug and make sure the lube is close to, or at, the level of the threads in the case. Unless there has been a leak, transmission fluid should be up to the level of the fill hole.

Add fluid, if needed, following the OEM procedures. Use the recommended lubricant as specified by the OEM. Some vehicles call for 75/80 or 80/90 weight lubrication oil, and one OEM calls for 30 weight motor oil in their manual transmissions. Check the specs to be sure.

Drive Shaft, Half-Shaft, and Universal Joints/Constant Velocity (CV) Joint (Front and Rear Wheel Drive)

6. Road test the vehicle to verify drive train noises and vibration.

The best way to determine is there is a problem in the drive train is to test drive the vehicle. Make sure there is no unusual vibration or noise when accelerating, decelerating or when turning on dry pavement at medium and very low speeds (~10 mph). Clicking or clunking noises may indicate worn U-joint or CV joint issues. Determine if any noticeable vibration stops when coasting at different speeds in neutral. This will help isolate tire and wheel issues related to imbalance, wear or defects from drive train concerns.

When a bearing is wearing out, it usually makes a clicking or howling noise as it rotates. The amount of noise depends on how much load is placed on it. Turns place additional thrust loads on wheel bearings so listen for front-wheel bearing noise when the vehicle is turning a corner.

A defective rear axle bearing noise is more noticeable at lower vehicle speeds when decelerating because there is less engine noise to mask it. You can diagnose rear axle bearing noise with the vehicle on a lift with the engine running and the transmission in drive. Run the vehicle at 35 to 45 miles per hour (56 to 72 kilometers per hour) and use a stethoscope placed on the rear axle housing directly over the axle bearings to listen for unusual noise. Noises, such as grinding or clicking, mean the bearings should be replaced. When a bearing fails, there is always a reason. The technician must correct the cause of failure to prevent the new bearing from failing.

7. Inspect, service, and replace shafts, yokes, boots, universal/CV joints; verify proper phasing.

When performing an undercar inspection on a RWD, AWD, or 4WD vehicle, check for loose driveshaft(s), universal joints, and yokes. Grasp each drive shaft and check for vertical movement in the universal joint. Try to rotate each drive shaft by hand and watch for movement between the driveshaft and the yoke. If vertical or rotary movement can be seen, replace the U-joint. Inspect the drive shaft for damage such as dents or missing balance weights.

For FWD vehicles, check the inner and outer CV joints for looseness or wear during the road test by listening for clicks during slow tight turns. Inspect inner and outer boots for cracks or leaks. Refer to Task B.4 for further information. Split-type replacement CV joint boots can be installed without removing the drive axles (half shafts).

Phasing of drive shafts involves making sure the universal joints are working in harmony. U-joints on the same driveshaft should be on the same plane, meaning they are in phase with each other.

When under the car, be sure to examine the CV joint boots for splits or cracks. Replace any boots that show signs of leakage ASAP to prevent possible CV joint damage.

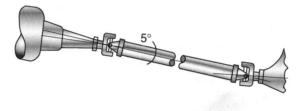

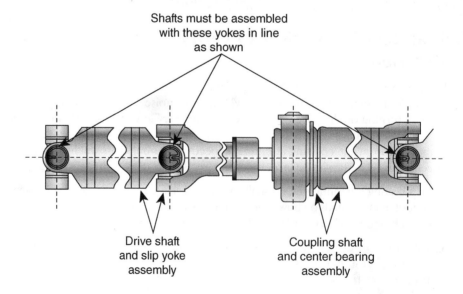

Shafts must be assembled
with these yokes in line
as shown

Drive shaft
and slip yoke
assembly

Coupling shaft
and center bearing
assembly

8. Inspect, service, and replace drive shaft center support bearings.

Some rear-wheel-drive vehicles have double (Cardan-type) universal joints mounted close together with a center yoke which connects the two universal joints and "splits the angle" between them. As described in Task C.7, check the U-joints and yoke for looseness. Replace any defective driveshaft support bearings or parts.

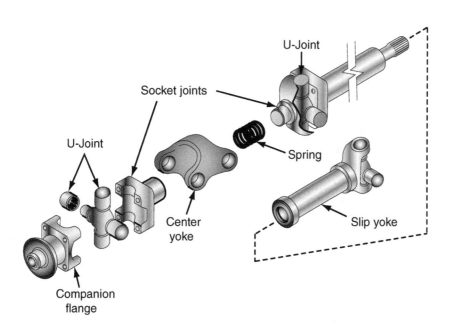

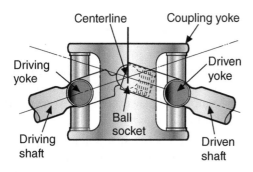

9. Inspect, service, and replace wheel bearings, seals, and hubs, excluding press-type bearings.

When a bearing fails, there is always a reason. The technician must correct the cause of a bearing failure to prevent the new bearing from failing. When a bearing is wearing out, it usually makes howling noise as it rotates. The amount of howling depends on how much load is placed on it.

A front wheel bearing usually provides a more noticeable howl when the vehicle is turning a corner because this places additional thrust load on the bearing. A defective rear axle bearing will usually make a howling noise that is more noticeable at lower vehicle speeds and when decelerating because there is less engine noise.

Diagnose rear axle bearing noise with the vehicle on a lift with the engine running and the transmission in drive. Run the vehicle at 35 to 45 miles per hour (56 to 72 kilometers per hour) and use a stethoscope placed on the rear axle housing directly over the axle bearings to listen for unusual noise. Noises, such as grinding or clicking, mean the bearings should be replaced.

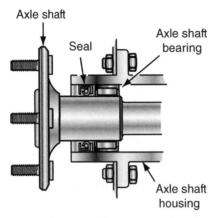

Rear axle rolling bearing and seal,
rear-wheel drive car.

Remove, clean, and inspect wheel bearings according to OEM maintenance schedules. Once removed, discard the used grease seal. Remove the bearings, wash them in solvent, and dry them with a lint free towel. Inspect the bearings for damage such as galling, abrasion, indentations, or other abnormalities.

If the bearings are in good shape, repack them with good quality wheel bearing grease. Avoid getting dirt or dust in them. Use a new grease seal whenever the bearings are being serviced or replaced.

If the wheel bearings are an integral assembly with the hub, you can check the bearing end play with the vehicle raised off the ground using a dial indicator. Place the indicator stem against the hub and move the hub in and out. If end play exceeds OEM specs, replace the hub and bearing assembly. Replace the bearing if it is noisy.

When front wheel bearings are mounted in the steering knuckle, inspect them for looseness. With the vehicle off the ground, grasp the wheel and attempt to wiggle it while feeling for play in the hub. If looseness is detected, check the bearings with a dial indicator, and discard them if they are out of specification.

TAPERED ROLLER BEARING DIAGNOSIS

Consider the following factors when diagnosing bearing condition:
1. General condition of all parts during disassembly and inspection.
2. Classify the failure with the aid of the illustrations.
3. Determine the cause.
4. Make all repairs following recommended procedures.

ABRASIVE STEP WEAR

Pattern on roller ends caused by fine abrasives. Clean all parts and housings, check seals and bearings, and replace if leaking, rough or noisy.

GALLING

Metal smears on roller ends due to overheating, lubricant failure, or overload. Replace bearing, check seals, and check for proper lubrication.

BENT CAGE

Cage damaged due to improper handling or tool usage. Replace bearing.

ABRASIVE ROLLER WEAR

Pattern on races and rollers caused by fine abrasives. Clean all parts and housings, check seals and bearings, and replace if leaking, rough or noisy.

ETCHING

Bearing surfaces appear gray or grayish black in color with related etching away of material usually at roller spacing. Replace bearings, check seals, and check for proper lubrication.

BENT CAGE

Cage damaged due to improper handling or tool usage. Replace bearing.

INDENTATIONS

Surface depressions on race and rollers caused by hard particles of foreign material. Clean all parts and housings. Check seals, and replace bearings if rough or noisy.

MISALIGNMENT

Outer race misalignment due to foreign object. Clean related parts and replace bearing. Make sure races are properly sealed.

Bearing failures and corrective procedures.

Rear Wheel Drive Axle Inspection

10. Identify fluid leakage problems.

On vehicles with a live (tubular) rear axle, lubricant for the bearings is provided by the lubricant in the differential. With the vehicle on a lift, carefully look for signs of oil or grease on the normally dry and dusty wheels and surrounding suspension components. Bearing failure on a live axle RWD vehicle may allow excessive lateral movement of the axle, which, in turn, would cause the axle seal to leak. When it does, lube oil leaking from the tubular axle housing will possibly contaminate the rear brakes. Also check for leaks at the differential pinion shaft seal and the differential cover. If leakage is evident, check if the differential has been overfilled or if the vent is clogged.

11. Inspect, drain, and refill with lubricant.

Remove the threaded inspection hole plug from the rear cover of the differential, insert your finger, and check for lubricant oil which should be close to the level of the threaded hole. If the oil level is low, try to determine the reason. There could be a leak, or perhaps oil has been thrown from the differential vent hole. If oil spills from the inspection hole, the differential may have been previously overfilled.

To change the differential gear lube, remove the drain plug and drain the old lubricant. If there is no drain plug, use an oil suction/extraction device. Insert the hose into the differential to draw out as much used gear oil as possible. It helps to do this with the differential at operating temperature. Once drained, install the drain plug and refill the differential to the level of the inspection hole using the OEM recommended type and viscosity of gear lube. Do not overfill the differential. Reinstall and torque the inspection plug to specs.

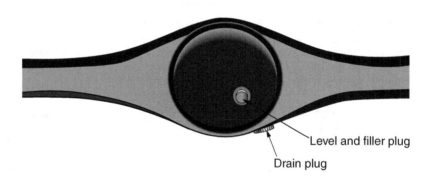

Level and filler plug

Drain plug

12. Inspect and replace rear axle shaft wheel studs.

With the rear wheel removed, inspect the condition of the axle flange and wheel studs that hold the wheel in place. Clean the flange with a wire brush to remove any debris or rust. Inspect the threads on the studs. If they are worn or damaged, replace them.

The studs are typically pressed/splined into position from the back of the flange. Install a spare lug nut loosely on a damaged stud and drive it out by striking the nut with a heavy hammer. Install a new replacement stud into the flange from the rear. Place some heavy washers on the stud and thread a good lug nut down onto the washer. Now use a wrench to pull the stud tight into position in the axle flange.

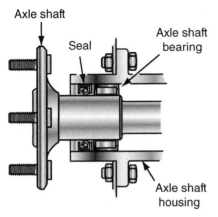

Rear axle with flange and studs,
axle bearing, and oil seal.

13. Inspect axle housing and vent; inspect rear axle mountings.

Check the axle housings for damage caused by striking rocks or road debris. Check the axle mounts and related hardware for damage or for being out of their normal position. Check the vent on top of the differential to make certain it is clear.

Four Wheel Drive

14. Inspect, adjust, and repair transfer case manual shifting mechanisms, bushings, mounts, levers, and brackets.

Due to the conditions under which 4WD vehicles operate, they need close periodic inspection of their undercar components. Rocks and other obstacles encountered while operating off-road can cause havoc with drive train components. The transfer case and related parts may be damaged even though they are typically protected by shields from impact by rocks, branches, and other debris.

Check that any undercar shields are where they should be and have not been removed. Look for unusual scrape marks or dents on the shields. Check for bent shift linkage, gearbox mounts and brackets or damaged bushings that would indicate severe off-road driving and the need for repairs.

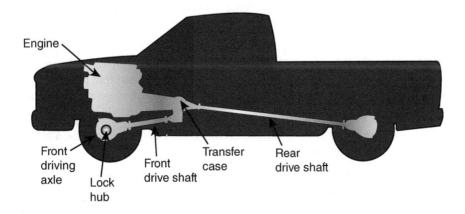

15. Check transfer case fluid level and inspect condition; drain and refill with fluid.

Remove the plug from the inspection hole in the transfer case and check for the proper level of gear lubricant. Refer to checking and replacing gear lubricant in the differential, as described in Task C.11.

16. Inspect, service, and replace front drive/propeller shaft and universal/CV joints.

A common complaint related to drive shafts and universal joints is a squeaking noise that increases in frequency with vehicle speed. Such noise is typically caused by dry or worn universal joints and is heard best while driving at low speeds. A worn universal joint may also make a "clank" noise when the transmission is shifted from park to drive or into reverse. A severely worn U-joint may emit a "clunking" noise at low speeds. Any of these noises calls for replacing the U-joints.

For troubleshooting CV joints, refer to Task C.7.

17. Inspect front drive axle universal/CV joints and drive/half shafts, axle seals, and vents.

The inspection of front drive U-joints and CV joints is similar to procedures for front- and rear-wheel-drive vehicles described in Task C.7. CV joint boots should be inspected regularly for cracks or tears. Check for loose or damaged clamps because these clamps keep the boots airtight. Check the boots and surrounding areas for splattered grease thrown from the boot.

Axle seals will eventually leak over time and should be periodically checked. If they are found to be leaking, determine the root cause as described in Task C.9.

18. Inspect front wheel bearings, seals, and hubs.

Whether used in front (FWD), rear (RWD) or 4-wheel drive (4WD) vehicles, bearings normally last a long time, so when they fail, look for the reason. Typical causes could be dirt ingestion caused by a faulty seal, water from the bearings being submerged (think boat trailer), or just dry from a lack of maintenance. The technician's job is to correct the root cause of a bearing failure to prevent the new bearing(s) from failing again.

When a bearing is on its way out, it usually makes howling noise as it rotates. The howling depends on how much load is placed on it. A front wheel bearing usually provides a more noticeable howl when the vehicle is turning a corner because this places additional thrust load on the bearing. Also, noises such as grinding or clicking means further investigation is needed.

For serviceable wheel bearings, remove, clean and inspect them according to OEM maintenance schedules. Discard the used grease seal once removed. Remove the bearings and wash them in solvent; dry them with shop air (but do not spin them!). Inspect the bearings for damage such as galling, abrasion, indentations, or other abnormalities.

If in good shape, repack the bearings with a good quality wheel bearing grease using a bearing packing tool if possible. Avoid getting dirt or dust in them. Use a new grease seal whenever the bearings are being serviced or replaced.

If the wheel bearings are an integral assembly within the hub, you can check the bearing end play with the vehicle raised off the ground. Using a dial indicator, place the indicator stem against the hub and move the hub in and out. If end play exceeds OEM specs, replace the hub and/or the bearing assembly as needed. Replace the bearing if it was noisy.

When front wheel bearings are mounted in the steering knuckle, inspect them for looseness. With the vehicle off the ground, grasp the wheel and attempt to wiggle it while feeling for play in the hub. If looseness is detected, check them with a dial indicator as described above, and discard them if out of spec.

19. Inspect transfer case, front differential, and axle seals and vents.

Check for leaks at the transfer case and front differential seals or gaskets. If an oil pan is used, some OEMs advocate re-using certain pan gaskets; some OEMs advocate using RTV sealant rather than a gasket. When using a sealant, use it sparingly making certain it is placed along the outside of the pan's bolt holes to minimize the possibility of sealant getting into the transfer case itself.

Check any axle seals for leakage of lubricant. If leaking, check for worn drive shaft/axle shaft bearings by pushing the drive shaft/axle shaft U-joint up, down and sideways while looking for lateral movement. If movement is noticed, the bearing may be worn. If so, it should be replaced along with a new seal.

Check the differential vent / breather located on top of the differential or possibly on the axle housing to make certain it is clear. If the vent is plugged, internal pressure buildup may force gear lubricant past solid-axle oil seals and contaminate the brakes.

20. Inspect tires for correct size for vehicle application; check for wear.

For safety reasons, and to ensure proper ABS, TPMS, and electronic stability control system operation, it is important to use OEM-recommended tire sizes on all four corners of the vehicle. If the vehicle owner has elected to plus size the tires or change wheel offset for any reason, the on-board electronic systems may not like the abnormal input signals associated with such modifications. Such modifications can result in the setting of fault codes, illuminated MILs, and a loss of vehicle handling and stability, which are usually provided by these vehicle features.

Likewise, uneven tire wear can cause a similar tripping of codes and the increased likelihood of handling or braking inefficiency, even on smooth, dry roads. 4WD vehicles especially need close tire inspection for damage caused by adventurous off-road driving, vehicle use at unpaved construction sites, or on unpaved washboarded and potholed roads.

Refer to the owner's manual (or specific tire manufacturer specifications) for proper tire size selection.

21. Retrieve and record diagnostic trouble codes (DTCs).

Part of the MLR technician's job is to understand how to use scan tools to retrieve and record diagnostic trouble codes stored in the vehicle's on-board computers. DTCs steer the technician in the right direction when diagnosing abnormal conditions in the vehicle. Once the codes have been retrieved and documented on the repair order, DO NOT erase them. Leave them for the technician assigned to doing the needed repairs to clear, once the vehicle faults have been corrected and the vehicle has been test driven and deemed correctly fixed.

> *Note:* Different manufacturers, and independent suppliers, offer scan tools with a whole variety of features.

While the OEMs provide scan tools that can read proprietary non-emissions-related DTCs, aftermarket scan tools provide features that may not be available on the OEM scan tools. In either case, select the scan tool that provides you with the needed diagnostic features. Scan tools range from simple code readers to bi-directional scan tools that read detailed sensor data (PIDs), put actuators through their paces, replay freeze frame data, access diagnostic and repair data wirelessly from the Internet, and serve to reflash PCMs and other on-board ECUs.

D. Suspension and Steering (13 questions)

1. Disarm airbag (SRS) system.

Service to the air bag system will require disarming/disabling the air bag deployment circuits; otherwise, accidental air bag deployment could occur, causing serious personal injury.

Check the OEM service manual and follow its procedures to the letter before performing any work on the Supplemental Restraint Systems (SRS) to avoid accidental deployment of the system. Technicians are too often injured by accidental air bag deployment.

In general, disabling the front and side (side curtain) air bags requires doing any of these procedures: turning off the ignition, removing the SRS fuse(s), disconnecting various connectors (yellow colored) behind the dash and under the seats. Be sure to wait the required time until the SRS module powers down (5 or more minutes).

2. Check power steering fluid level; determine fluid type and adjust fluid level; identify system type (electric or hydraulic).

Most vehicles use hydraulic power steering systems. Look for a belt-driven power steering pump and reservoir under the hood. Some GM vehicles have a combined hydraulic power steering and power brake systems (called HydroBoost). Some contemporary vehicles are using Electronic Power Steering (EPS) and Variable Effort Steering (VES) systems. Such systems rely on the Vehicle Speed Sensor (VSS) as input to a steering controller that in turn modulates steering pressure according to vehicle speed.

Checking the level of hydraulic power steering (PS) fluid is a regular maintenance item, even though power steering fluid does not normally need to be "topped off." Some vehicle manufacturers recommend checking the fluid level when it is up to operating temperature and the engine is running. Normally, a dipstick on the pump reservoir is used to check the fluid level. To check the fluid level, remove the dipstick, wipe it clean with a lint-free rag or towel, reinsert it, and withdraw it to read the fluid level on the dipstick.

There are different formulas of PS fluid being used in various vehicles, so when adding or replacing PS fluid, be sure to only use the OEM-recommended type of fluid.

Hybrid and battery electric vehicles use electric powered rack and pinion steering for less engine parasitic load (from a PS pump) and improved gas mileage. These systems do need or have a hydraulic PS pump. Watch for an increase of electric power steering systems as new vehicle models appear on the market.

3. Inspect, adjust, and replace power steering pump belt(s), tensioners and pulleys; verify pulley alignment.

Part of a routine maintenance inspection should include checking the power steering pump belt condition and its tension. Check for a shiny belt surface, cracks, splits, and other defects on V belts. On serpentine bests, look for wear using an approved wear gauge.

Some pulleys like this alternator decoupler pulley (ADP) are special because they 'synchronize' the belt drive system for improved engine efficiency to reduce Noise, Vibration and Harshness (NVH) and increase component life. ADPs are not interchangeable, so be sure to use the correct replacement part.

Check belt tension with the engine OFF by pressing down on it between two pulleys. Take note of how much the belt deflects and compare the deflection to specifications. You can also use a belt tension gauge. The belt should be tight, but not too tight. Any squealing when the steering wheel is turned against the left or right lock indicates the belt is too loose or is dried out and needs replacing.

Belts are kept tight as they wear by belt tensioners. Some tensioners have wear indicators to show if the belt has stretched or become worn beyond its service life. Tensioner bearings will make noise as they become faulty and need replacing. Listen for louder than normal "running noise" or whining from the tensioner. Use a piece of hose or tubing as a stethoscope to localize belt tensioner or pulley noises, being ever mindful of possible entanglement with moving engine parts.

A variety of tensioner types are used on contemporary vehicles. Some overrun to reduce parasitic engine losses, improve vehicle mileage, and reduce emissions. Some even help smooth out torque reversals from the crankshaft pulley, which would affect the belt-driven accessories. Make certain the correct style of tensioner is installed when replacing the belt tensioner.

A serpentine (ribbed V) belt routing diagram sticker should be affixed to the radiator cowling or on the underside of the hood. Check that the ribbed portion of the belt is properly seated in all pulleys and not offset by one or more ribs. With the engine OFF, check for pulley alignment using a straightedge. Any misalignment must be corrected by replacing defective parts or by realignment of accessory mountings. Serpentine belt manufacturers tell us that modern belt composition does not show the usual signs of dryness and cracking as they age. The proper way to test serpentine belts is by using a wear gauge (see image).

4. Identify power steering pump noises, vibration, and fluid leakage.

If the PS fluid is low, a groaning noise in the power steering pump may be heard, or hesitation at the steering wheel may be felt. With the system up to the proper level of fluid, turning the steering wheel from "lock to lock" should be smooth and free of noise. Expect the pressure relief valve to make a hissing sound when the steering wheel reaches the end of its travel; do not hold the steering wheel in this position for more than a few seconds.

The steering wheel should rotate with equal pressure towards both the left- and right-hand direction. Use a spring scale on the steering wheel to be certain. You can also check power steering pump pressure when turning the steering wheel back and forth. Use a pressure gauge and "T" plumbed into the high-pressure power steering hose. Any imbalance of pressure between left and right turns could be caused by defective seals in the power steering components.

Inspect the entire power steering system for leaks. Start at the pump and reservoir, and then inspect the PS rack and high-pressure hoses. Be sure to include checking the PS fluid return line leading back to the reservoir for leaks, and for signs of power steering fluid on or around the PS rack bellows.

5. Remove and replace power steering pump; inspect pump mounting and attaching brackets; remove and replace power steering pump pulley; transfer related components.

Inspect power steering pump mountings with the engine running. Have an assistant turn the steering wheel about one- half turn in both directions. Watch for any unusual movement of the steering pump as the wheel is turned. Any pump movement in its brackets or bushings means the brackets need tightening or the bushings need to be replaced.

When replacing a power steering pump, loosen the hoses and catch the power steering fluid in a suitable pan or container. Once the pump has been removed, take off its pulley, mounting brackets, and associated hardware and transfer these items to the replacement pump before installation. Start the engine and check for leaks.

6. Inspect and replace power steering hoses, fittings, O-rings, coolers, and filters.

A thorough underhood and undercar inspection for leaks in power steering hoses and fittings should be performed whenever a vehicle is brought in for routine service. Power steering high-pressure hoses may have become old and hardened, or perhaps the "O" rings used in PS components have suffered from use or aging. In either case, looks for telltale leaks that indicate the need for service.

If used, check the undercar cooler for leaks caused by stones or road debris. If there is a PS filter, be sure to change it according to the respective vehicle maintenance schedules. It may be a separate in-line filter, or it may be located at the bottom of the PS fluid reservoir.

7. Inspect and replace rack and pinion steering gear bellows/boots.

Squish the PS bellows (boots) on either side to determine if power steering fluid has accumulated inside. Hard to squish stiff-type bellows will need to be slid back to check for rack seal leakage as evidenced by PS fluid buildup inside. Carefully inspect the bellows for dryness and for cracks, tears, or other defects. If loose, but otherwise serviceable, tighten the clamps to the manufacturer's specifications. If a boot needs replacing, slice it open to remove it, and at the same time, inspect the rack seals for leakage. If rack seal leakage is noted, the rack will need to be removed and replaced. To replace a bellows, the tie rod end must normally be disconnected and removed in order to slide a new bellows into place.

8. Flush, fill, and bleed power steering system.

If the power brake reservoir is low on fluid, find out why. Its level does not drop unless there is a leak. Likewise, if the fluid is discolored (brownish in color) or contaminated with moisture, rust, or dirt, it is time to drain and flush the system and replace the old PS fluid with fresh PS fluid. Even if it the fluid appears to be clean, check the maintenance schedule as well to see if it is time for such service.

Be sure to check the manufacturer's service procedures for the correct PS system drain and flush procedure. A general procedure involves placing the vehicle on jack stands, removing the PS return line, starting and idling the engine, and letting the PS fluid drain into a suitable pan. At the same time, add fluid periodically until the fluid coming out looks as fresh as the fluid being poured in. Stop the engine when this point is reached, and replace the drain line. Fill the reservoir with fresh fluid.

To bleed the system, start the engine and move the steering wheel from lock to lock repeatedly. Add more fluid to the reservoir as required until no "growling" noises are heard in the system and the steering wheel turns smoothly with no jerkiness. Once the fluid is hot, top off the reservoir to the full mark.

Steering Linkage

9. Inspect, adjust (where applicable), and replace pitman arm, center link (relay rod/drag link/intermediate rod), idler arm(s) and mountings.

On older RWD vehicles, traditional recirculating ball steering gears and parallelogram steering linkages were widely used. Recirculating ball steering gears are typically bolted to the vehicle frame and these bolts must be checked for tightness. If necessary, torque them to manufacturer's specifications.

All links, rods, and rod ends used in parallelogram steering systems must be inspected for being bent or loose. Grasp each component and shake it to see if there is any play in the steering linkage. Likewise, with the front end on jack stands, start the engine. With the engine idling, have an associate move the steering wheel from lock to lock as you watch beneath the vehicle for any unusual movement or play in the connecting links and/or mountings.

10. Inspect, replace, and adjust tie rods, tie rod sleeves/adjusters, clamps, and tie rod ends (sockets/bushings).

Tie rod ends take a beating and need to be routinely inspected for wear. Collisions with rocks and road debris can damage low-hanging steering components, including the tie rod boots, adjustment sleeves, connecting links, and the tie rod ends themselves.

To check the tie rod ends with the front wheels off the ground, grasp each front tire at the 3 o'clock and 9 o'clock positions and try to rock the tire sideways. Any looseness could be due to play in the parallelogram steering linkage, but more likely it is due to the typically beaten and worn tie rod ends or steering knuckles. Some tie rod ends have wear indicators, or markings, that serve to indicate if the tie rod end is worn beyond its safe service life.

You can also check for steering system looseness from behind the wheel. With the front wheels straight ahead and the engine off, rock the steering wheel gently back and forth from straight ahead and take note of the steering wheel's free play; compare it to specifications. If there is excessive play, check all of the steering components for looseness, wear, or damage. If the tie rods are replaced, the toe adjustment is upset and a front-end alignment is required.

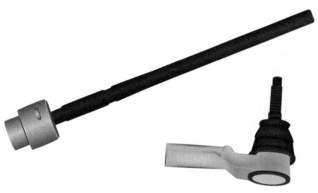

Tie rod ends take a beating and need to be periodically inspected and replaced.

11. Inspect and replace steering linkage damper(s).

Along with the venerable air-cooled VW bug, some rear wheel drive luxury vehicles, trucks and SUV-type vehicles, use a steering damper to help reduce kickback to the steering wheel when rocks or potholes are encountered. Steering dampers act like horizontal steering linkage shock absorbers, and they are checked similarly to suspension struts/shocks for damage, wear, leakage, or looseness. In order to check steering dampers, drop one end of the damper and flex the rod in and out. The same resistance should be encountered when the rod is moved in either direction. If the damper rod does not offer enough resistance to endwise movement, replace the damper.

Front Suspension

12. Identify front suspension system noises, handling, ride height, and ride quality concerns; disable air suspension system.

There are a number of possible causes for suspension system noise complaints; for example, worn tie rod bushings may cause a rattling noise on road irregularities. Worn 4WD front leaf spring shackles and bushings may also cause a rattling noise on rough or uneven roads. Worn sway bar bushings may be the cause of a knocking sound when driving slowly over irregular roads or driveway surfaces. Dried-out sway bar bushings may squeak or emit a knocking sound. Upper strut bearings on MacPherson strut suspensions may knock when the vehicle is driven over uneven or unpaved surfaces.

Improper vehicle handling or improper ride height of the vehicle may be caused simply by incorrect loading of the vehicle (excessive or unbalanced loading). Vehicle handling concerns may also be due to incorrect wheel alignment settings, or because of worn tires or shocks/struts. Incorrect handling could also be caused by worn or damaged steering and suspension components themselves.

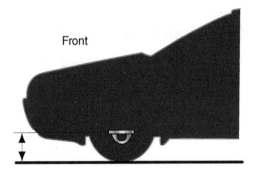

Curb riding height measurement, front suspension.

Check the service manual for proper ride height specifications. Check the vehicle with the tires properly inflated and the normally carried provisions on board. If the ride height is too low, suspect worn or collapsed springs or an excessive load on the vehicle.

> **Note:** Whenever a vehicle with air suspension is to be raised on a floor jack or lifted off the ground, be sure to disable the air suspension system by first locating and turning OFF the on-board air pump. Follow the vehicle manufacturer's instructions when disabling air suspension systems to avoid burning out the pump from overwork.

13. Inspect upper and lower control arms, bushings, and shafts.

Worn upper and lower control arm bushings affect the front end caster and camber angles (see Wheel Alignment Tasks D.41 and D.42) of the front suspension. If incorrect, these adjustments may cause vehicle pulling to one side or the other, along with excessive tire wear. Check the upper and lower control arms for any loose conditions. If the control arm bushings are worn, erratic steering and/or noises may be heard by the vehicle driver when driving on uneven or irregular roads. Dry or worn bushings may squeak as the vehicle bounces up and down.

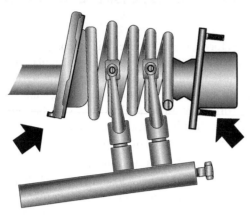

Aligning lowest bolt on upper strut mount with tab on lower spring seat.

14. Inspect and replace rebound and jounce bumpers.

Rebound bumpers should be inspected for cracks, wear, or damage. Sagging springs and low ride height may be the cause of damaged bumpers. Also check for worn shocks or struts.

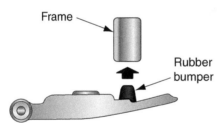

Rebound bumper.

15. Inspect track bar, strut rods/radius arms, and related mounts and bushings.

A track bar (aka tracking bar or panhard rod) helps to stabilize the chassis from lateral movement. Strut rods prevent fore-and-aft movement of the lower control arms. A bent or loose strut rod or tracking bar or loose, worn, or damaged bushings or mountings could contribute to handing (pulling) and ride problems. Check that these parts are firmly in place and free of defects.

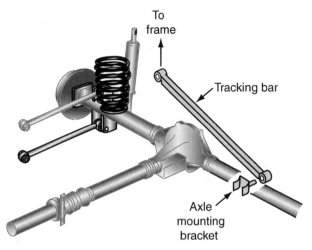

Tracking bar.

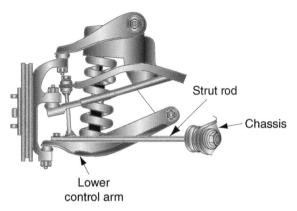

A strut rod.

16. Inspect upper and lower ball joints (with or without wear indicators).

To check the ball joints, raise the vehicle front end about six inches off the ground to unload the tire. Slide a large tire iron under the tire and pry it up and down. As you do, feel for looseness in the ball joints. With the wheels raised, you can also grasp the top and bottom of the tire and push/pull on it while watching and feeling for looseness. When under the car's front end, wiggle the tire in and out while watching for play in the ball joints.

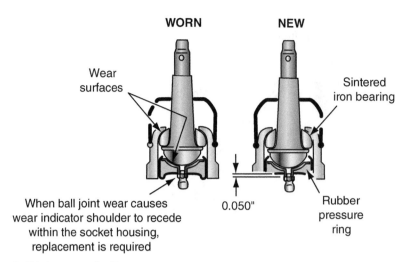

Wear surfaces

Sintered iron bearing

When ball joint wear causes wear indicator shoulder to recede within the socket housing, replacement is required

0.050"

Rubber pressure ring

Ball joint wear indicator.

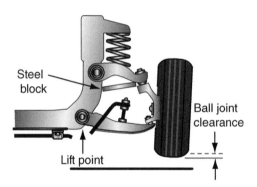

Steel block

Ball joint clearance

Lift point

Live-axle, leaf spring rear suspension system.

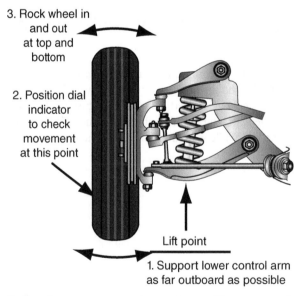

3. Rock wheel in and out at top and bottom

2. Position dial indicator to check movement at this point

Lift point

1. Support lower control arm as far outboard as possible

Coil-spring rear suspension system with upper and lower control rods.

17. Inspect non-independent front axle assembly for damage and misalignment.

Non-independent front ends are designed for rough vehicle use, but they can be knocked out of alignment when off-roading or undergoing similar abuse. Some SUVs or pickups use a single I-beam; others may use a twin I-beam axle setup. Abnormal tire wear is an indication that the I-beam may be bent or otherwise out of alignment. Look for signs of rock impact (where rust has been scraped off) or where front-end repairs to the vehicle (fresh paint or new parts) has been performed, which could mean an alignment is required.

18. Inspect front steering knuckle/spindle assemblies and steering arms.

With the vehicle on a lift, check the steering knuckle and tie rods for looseness. Tie rods must be tight on their respective steering knuckles, and they must be secured with a cotter pin. On parallelogram-type steering, check for loose steering arms and inner and outer tie rod ends by firmly grasping the steering arms and wiggling them while watching for any looseness or feeling of play.

19. Inspect front suspension system coil springs and spring insulators (silencers).

Coil springs may show signs of sagging on older vehicles. Check ride height to verify coil spring condition. Look for any collapsed coils or signs of abrasion from road debris. Bounce the vehicle while watching and listening for binding or noises. If used, make sure the spring silencers are intact and serviceable.

20. Inspect front suspension system leaf spring(s), leaf spring insulators (silencers), shackles, brackets, bushings, center pins/bolts, and mounts.

Front-end alignment and tire wear can be affected by weak or broken leaf springs. Such a condition can also reduce directional stability and cause a rough ride. Worn leaf spring shackles and bushings might also affect the vehicle's ride height. Inspect and check for looseness of these parts by firmly grasping them and shaking them. Replace any worn of damaged components or parts.

21. Inspect front suspension system torsion bars and mounts.

If equipped with them, check the vehicle's front torsion bars for straightness and for loose or broken mounts. If any damage is noted, replace the failed parts.

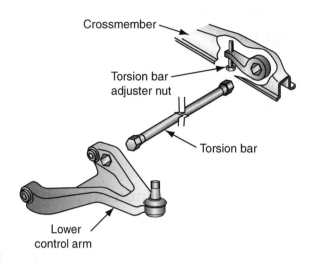

Crossmember

Torsion bar
adjuster nut

Torsion bar

Lower
control arm

22. Inspect and replace front stabilizer bar (sway bar) bushings, brackets, and links.

A weak stabilizer (sway) bar or worn bushings will contribute to vehicle sway when going around turns or to vehicle stability when driving over rough roads. When driving slowly over irregular surfaces, listen for a knocking noise which may indicate worn sway bar bushings. In rare instances, the sway bar itself may become cracked or broken. All mountings should be visually inspected for wear or deterioration. Some vehicles have sway bar links. Check for bent or loose links and end connections.

23. Inspect front strut cartridge or assembly.

Check the front struts for weakness by pushing down on each front corner of the vehicle. The vehicle should not continue to bounce or rock after letting go. Check for rust, damage, and for oil leakage from the struts. If any of these conditions exist, replace both front struts as a pair. After replacement, check the front-end alignment.

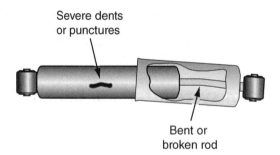

Severe dents
or punctures

Bent or
broken rod

24. Inspect front strut bearing and mount.

Strut bearings are critical for smooth steering wheel operation, as well as for front-end alignment. Check that excessive effort is not required to turn the steering wheel. Check also for looseness at the upper end of the struts and listen for knocking noises when going over bumps.

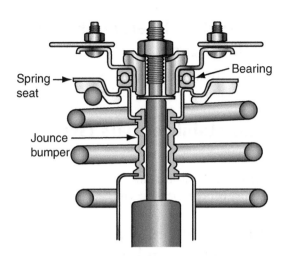

25. Identify noise and service front wheel bearings/hub assemblies.

Wheel bearing noise is most evident when the bearings are side loaded during turns. When turning one direction and then the other at low speeds, listen for an unusual "growling" noise coming from the bearing area.

Many wheel bearings are now "sealed" (no service possible) and pressed into place in the front hubs. On older vehicles, (plus small trailers), ball and roller wheel bearings must be removed, cleaned, and repacked with bearing grease according to the maintenance schedule. If they are subjected to severe duty (e.g., exposed to heavy dust or sand; submerged into water at boat ramps, etc.), they will require more frequent service. Inspect these bearings for galling, pitting, or brinelling, and replace them as required along with a replacement grease seal.

Rear Suspension

26. Identify rear suspension system noises, handling, and ride height concerns; disable air suspension system.

Noises from the rear of the vehicle could range from squeaking to growling. Perform a test drive with the windows down, listening for gear noise from the RWD differential, an axle bearing growl from a non-independent/solid rear axle assembly, or noises from dry or damaged wheel bearings. Check for vehicle sway during turns, for a harsh ride, or for sagging or squeaky springs.

Check the service manual for instructions on how to disable the air suspension system before raising the vehicle off the floor (so the air springs' pump does not continue running and become damaged).

27. Inspect rear suspension system coil springs and spring insulators (silencers).

Check the coil springs for collapsed coils or excessive rust. Check the spring seats / insulators for damage which could allow noise to be transferred to the vehicle body. Check the vehicle's ride height.

28. Inspect rear suspension system lateral links/arms (track bars), and control (trailing) arms.

Check that the track bar (aka Panhard rod) and control arms (if used) and bushings are in good shape, which would help prevent unstable vehicle handling.

29. Inspect and replace rear stabilizer bars (sway bars), bushings, and links.

Check the condition of the (anti) sway bar and its bushings for dryness or cracking. If used, check the sway bar links for being bent or for loose ends.

30. Inspect rear suspension system leaf spring(s), leaf spring insulators (silencers), shackles, brackets, bushings, center pins/bolts, and mounts.

Leaf springs should be inspected for sagging, which would affect ride height, causing it to be less than specified. Visually inspect the individual leafs for cracks. The entire spring or individual leafs should be replaced if the springs are sagging or broken. Check for broken center bolts, worn shackles, or bushings.

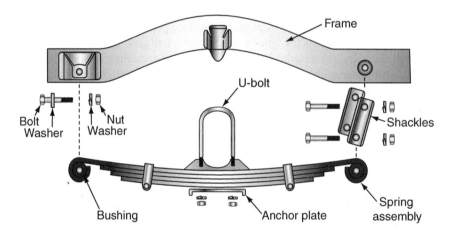

Check the plastic silencers between the spring leafs. If worn out, creaking and squawking noises are heard when the vehicle is driven over road irregularities at low speeds. When checking or replacing silencers, raise the vehicle with a floor jack and place jack stands under the frame rails so the rear suspension moves downward and the spring spreads apart. With vehicle weight removed from the springs, the leafs can be pried apart to remove and replace the silencers.

Worn spring shackle bushings, brackets, and mounts will likely allow unnecessary chassis lateral movement and produce rattling noises. With the vehicle weight on the springs, insert a pry bar between the rear outer end of the spring and the frame rail. Push down on the bar to check the rear shackle for movement. Shackle bushings, brackets, and mounts should be replaced if there is movement in the shackle.

A broken spring center bolt could allow the rear axle assembly to suddenly shift rearward on one side, which would alter the rear wheels' tracking (a frightening experience!). This severely affects handling, tire wear, and reduces directional stability.

31. Inspect and replace rear rebound and jounce bumpers.

Rear rebound bumpers are normally bolted to the chassis. Check for cracks or flattened bumpers. Rebound bumper damage is usually caused by weak and sagging springs or by worn out shock absorbers or struts. Replace any worn or damaged bumpers as required.

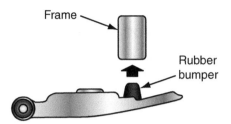

32. Inspect rear strut cartridge or assembly, and upper mount assembly.

Check the rear struts for oil leakage or damage. If either is evident, replace both rear struts as a pair. Check the upper strut mountings for damage caused by heavy impact from rough roads and/or heavy vehicle loads.

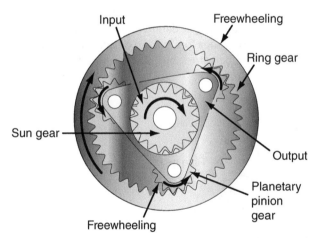

Shock absorbers are another part that take a beating. Check them for rust, cracks, oil leakage and dry or deformed bushings.

33. Inspect non-independent rear axle assembly for damage and misalignment.

The rear axle can be shifted out of proper alignment due to hard impact with pot holes or road objects, or if the vehicle has been improperly lifted or towed. Axle housing misalignment will cause "dog tracking" of the vehicle. Check for damaged U-bolts and mountings of leaf springs and other components. Look for obvious signs of impact damage. Performing a 4-wheel alignment on the vehicle will reveal if the rear axle has shifted.

34. Inspect rear ball joints.

In independent suspension vehicles, check the rear ball joints for looseness or for a lack of lubrication. With the vehicle off the ground, grasp the tire and wiggle it from the 6 o'clock and 12 o'clock positions while observing the ball joints. Any ball joint play would indicate the need for replacement.

35. Inspect and replace rear tie rod/toe linkages.

Check the rear tie rods for looseness or for indicated wear; check the grommets and mountings. Replace components as needed and tighten to specifications. When these parts are replaced, be sure to check the rear wheel toe settings.

36. Inspect rear knuckle/spindle assembly.

With the vehicle raised, watch for any knuckle play by grasping the tire and firmly wiggling it. With the rear wheel, brake drum, and wheel bearings removed, check the spindle for damage or being bent out of shape. Check for galling or rough spots that would prevent the bearing races from properly seating.

37. Inspect and replace shock absorbers, mounts, and bushings.

Check the shocks for damage, rusting, or oil leakage (a slight coating of oil mist is permissible). Bounce the car to see if the shocks dampen movement with one rebound. Check the upper and lower shock mountings for damage caused by vehicle overloading or from weak springs. Check that the rubber shock bushings are not dry or have been squeezed part way out of the mountings.

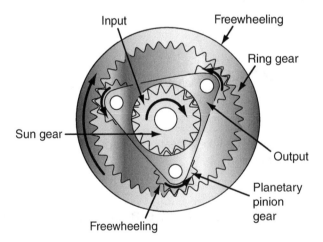

38. Identify noise and service rear wheel bearings/hub assemblies.

Drive the vehicle slowly and listen for a clicking noise, or in extreme cases, a grinding or growling sound, from the rear wheel bearings. If any noises are heard, disassemble serviceable rear wheel bearings and inspect for damage. If the bearings are sealed and pressed into a hub, press them out and replace them as a unit.

Wheel Alignment

39. Identify alignment-related symptoms such as vehicle wander, drift, and pull.

A recommended way to perform an alignment-related diagnosis is to take the vehicle on a test drive on a flat and level road and note any handling irregularities. Notice if the steering wheel is off-center as the vehicle travels straight ahead or if it has excessive play. Does the vehicle drift to one side of the road if the steering wheel is not held, and if so, towards which side? Does the vehicle shimmy, tend to wander within the lane when the steering wheel is pointed straight ahead, or experience "nibble" (slight steering wheel oscillations)? Does hitting irregularities in the road cause the vehicle to dart towards one side of the road or the other (bump steer)? Enter all such symptoms on the repair order and make certain the appropriate repairs are performed by a properly trained front-end specialist. Following any such repairs, a 4-wheel alignment will be called for.

40. Perform pre-alignment inspection; prepare vehicle for alignment, and perform initial wheel alignment measurements.

Make sure the tires are inflated properly. With the vehicle on the alignment rack, examine all four tires carefully for any clues indicating incorrect inflation, misalignment (see Task D.48). Make sure the tires match, i.e. are of the same type, correct size, and are mounted on the correct rims. If the tires are directional or asymmetrical, make sure they are properly positioned on the vehicle. Check to see if changes have been made to "plus size" the tires, or if rim offset has been changed from OEM specifications. Such changes will affect road feel, steering response, and handling.

41. Measure front and rear wheel camber; adjust as needed.

As viewed from the front of the vehicle, camber angle is the amount the wheel "leans" inward or outward from top to bottom. The camber angle may be positive (top leans outward), negative (top leans inward), or neutral (straight up and down). Camber affects the vehicle's ability to track properly and not dart from one side or the other. Check for tire shoulder wear, which is indicative of incorrect camber. Check the vehicle specifications against actual readings done on the alignment machine and correct the camber as required.

42. Measure caster; adjust as needed.

The caster angle is the amount of forward or rearward "lean" of a wheel's axis as viewed from the side of the wheel. If caster is "positive," the upper suspension pivot (ball joint or strut bearing) is rearward of the lower pivot. If caster is "negative," the upper pivot lies forward of the lower pivot. Most vehicles are designed with a slight amount of positive caster (think of a bicycle), but they may have zero or negative caster, as measured in degrees. Caster affects the ease of vehicle turning and the return of the vehicle's steering wheel to center. Thus, caster directly affects directional stability and tire wear. Front caster may not be adjustable on MacPherson strut front suspensions without fitting the OEM strut bearing with adjustable strut mounts. Set caster to specifications.

43. Measure front wheel toe; adjust as needed.

The vehicle's "toe-in" or "toe-out" is measured in either inch increments or in angles in degrees. Considering the two tires at the same axle (front or rear), the distance between the front of the tires is compared to the distance between the rear of the tires. With the steering wheel straight ahead, toe-in means the fronts of the two tires are closer to each other than the rear, and vice versa. If toe is incorrectly set, tire wear (feathering of the tread), turning response, and vehicle handling is affected. The wheels will toe-out when turning with the two front wheels assuming different degrees of toe. This helps to decrease the turning radius of the vehicle and makes turning easier with no tire "scrubbing."

The vehicle's toe angle also changes as the vehicle is propelled straight ahead. The OEM specifications may call for positive, negative, or neutral toe settings so that the wheels are effectively straight ahead when the vehicle is typically loaded and underway.

All vehicles have front-wheel toe adjustments; independent rear suspension vehicles provide for rear-wheel toe adjustments. Toe adjustments on the front wheels are made at the tie rod end adjustment sleeves. Adjust to specifications.

44. Center the steering wheel using mechanical methods.

In the interest of customer satisfaction, it is important to center and affix the steering wheel in the straight ahead position before starting a wheel alignment procedure. A special steering wheel holder (see image) is used to secure the steering wheel when certain alignment adjustments (ex: toe in) are being performed.

Use a tool like this to keep the steering wheel centered when performing a wheel alignment.

45. Measure rear wheel toe; adjust as needed.

No OEM rear toe adjustment is provided on vehicles with solid rear axles, but live (independent) rear suspension vehicles have provisions for rear -toe adjustments. Correct rear wheel toe is critical to vehicle stability when turning to help avoid "spinning out," especially under marginal road conditions.

46. Measure thrust angle.

The vehicle's thrust angle is defined as the actual direction in which the rear wheels point as compared to the geometric centerline of the vehicle. The positioning of the rear axle and rear wheel alignment affects the thrust angle and must be considered when performing a 4-wheel thrust line alignment. The optimum vehicle thrust angle is zero degrees.

47. Calibrate steering angle sensor.

After an alignment has been performed on a vehicle equipped with an electronic stability control (ESC) system, check that the steering angle sensor (which provides input to the ECU) is properly calibrated. Check that with the steering wheel pointed straight ahead there are no DTCs logged and no illuminated MILs indicating the steering angle sensor needs to be "zeroed." If necessary, connect a scan tool and perform a recalibration procedure according to the OEM procedures.

Wheel and Tires

48. Identify tire wear patterns.

Look for worn outer or inner shoulders that indicate incorrect camber. Check for signs of toe-related feathering (sawtoothed tread), scrubbing, tire-imbalance related cupping, or other abnormal tire wear patterns.

49. Inspect tire condition, tread depth, size, and application (load and speed ratings).

Check for any bulging of the sidewalls. Check for tears or splitting in the tread area. Make certain the tires are not worn down to the wear bar indicators, or use a tread-depth gauge to determine remaining tire life. Check with the tire manufacturer's resource manual to make certain the tires are the correct type, size, and speed rating for the vehicle, and that they are not overloaded. Some vehicles require a different tire size at the front versus the rear of the vehicle.

Use a tire tread-wear gauge like
this to determine the remaining
useful life of a tire.

When inflating tires on a vehicle, use the vehicle
manufacturers tire pressure information found on
the door or door jamb. The maximum inflation
pressure indicated here is for general purposes only.

50. Check and adjust tire air pressure. Utilize vehicle tire placard and information.

Use an approved tire pressure gauge (digital readout or "stick" type) to make certain the
tire pressure, as indicated on the driver's door or doorjamb sticker, is correct at all wheels.
Some vehicles require a different inflation pressure at the front tires versus the rear tires of

the vehicle. Do not use the maximum inflation pressure spec on the tire itself for inflating the tires for a specific vehicle being serviced, as the max. pressure spec is likely too high.

51. Diagnose wheel/tire vibration, shimmy, and noise concerns; determine needed repairs.

Check tires for cupping or bulging, which may cause vehicle vibration or steering wheel pulsations and noise. Unusually loud tire noise at higher speeds could be caused by harmonics generated in the vehicle's suspension system. Check the OEM's TSBs for recommendations of a change of tire manufacturer and/or tire type and model for a quieter ride.

52. Rotate tires/wheels and torque fasteners/wheel locks.

Check that wheels, studs, mounting faces, and bolt hole chamfers are clean and free of dirt, rust, debris, or wear. Do not lubricate the studs or wheel nuts. Rotate the tires per the OEM maintenance schedule taking into account the type of tire (bias ply or radial) being installed and the type of vehicle drive (RWD and 4WD versus FWD). Make certain that directional and asymmetrical tires are mounted on the correct side of the vehicle. Inspect, install by hand, and lightly wrench tighten wheel mounting bolts or lug nuts. Torque them to factory specifications in an alternating sequence. After tightening locking type lug nuts or bolts, be sure to return the key to the vehicle owner.

53. Dismount and mount tire on wheel.

When dismounting tires from their rims, follow the tire changer machine manufacturer's instructions. To avoid damage to TPMS sensors, start separating the bead from the rim with the valve stem at the OEM-required position (i.e., some recommend breaking the bead at 6 o'clock, some at 9 o'clock, from the valve stem position). Do not remove the valve stem entirely; if it needs replacing, be sure to use the properly plated valve cap and stem.

When remounting the tire on the rim, follow the tire machine manufacturer's instructions along with OEM guidelines. Make certain the tire and the rim match regarding their size and bead taper angle. Replace the rubber tire valves used on conventional wheels.

54. Balance wheel and tire assembly.

Except for emergencies, avoid static tire balancing. According to tire manufacturers, the most desirable method of balancing a tire and wheel assembly is by using an off-the-car computerized tire balancing machine. Follow the tire balancing machine's manufacturer's instructions for accurate dynamic balancing.

55. Identify and test tire pressure monitoring systems (TPMS) (indirect and direct) for operation. Verify instrument panel lamps operation; conduct relearn procedure.

Various automobiles make use of different types of TPMS systems. Look up the system used for the vehicle in question in the vehicle service manual or the tire manufacturer's Tire Resource Manual. Check that the TPMS dash light goes out with the key ON and the tires are properly inflated. When rotating tires, a TPMS relearn/reset procedure should be performed to ensure proper location ID of a low tire pressure situation. A scan tool or special TPMS device may be needed to do this. TPMS Relearn is done with the ignition ON and in a particular tire location sequence (e.g., LF, RF, RR, LR). DO NOT forget to include the spare tire in this process.

56. Repair tire according to tire manufacturers' standards.

Tire repair should only be performed to the tread area of the tire, not to the sidewall. The puncture or injury should not be larger than ¼ inch (6 mm) in diameter. The tire should be removed from the rim in order for a proper tire inspection and repair to be performed. The one-piece tire plug and patch is the preferred (and fully legal) way to repair a hole in a tire. Follow instructions that are provided with the patch kit, making sure the repair area is properly cleaned, buffed, vulcanized, and ultimately sealed.

E. Brakes (11 questions)

1. Check for poor stopping, pulling, dragging, noises, high or low pedal, and hard or spongy pedal.

Before taking a vehicle with a braking system complaint for a test drive, make absolutely certain to read the repair order, looking for any indication of a brake system failure. When safe to do so, drive the vehicle slowly, away from other traffic, and apply the brakes lightly. Take careful note of the vehicle's brake system behavior.

Listen for noises such as the grinding of rivets against the brake drums or of disc brake backing plates grinding against the rotors. Listen for the squeal of a wear indicator against a rotor, especially with the brakes applied and during turns.

Check for pulling to one side or dragging that could indicate something is broken at one wheel or that a brake line is pinched of clogged. A pulsating pedal under moderate or severe braking may be caused by one or more warped rotors. Check if the power-assisted brakes are hard to apply, the vacuum booster may have failed or there is a vacuum line leak. There could also be a partial system failure or pinched or clogged brake lines.

Take note of the brake pedal height. If it is high (less than about a quarter inch), check the vehicle specs. The pushrod may be too long and it needs to be shortened. If the pedal is low (too much free play) it may be due to brake wear, damaged parts, a brake fluid leak, or a too short pushrod. If it goes to near the floor, the system may be leaking brake fluid from one circuit. If it goes completely to the floor (in a dual circuit brake system), there is likely a serious mechanical failure. It is unlikely that both circuits have failed. Take note if the pedal is soft or "spongy," which may indicate the need for bleeding or the need to replace the seals. Seal failure could be due to wear and tear or contamination by petroleum-based liquids.

Note if the pedal sinks slowly when lightly applying and holding the brakes while stopped; a seal in the master cylinder may be faulty, requiring the master cylinder to be rebuilt or replaced.

2. Check the master cylinder fluid level and condition; inspect for external fluid leakage.

One of the first things to check during a maintenance inspection is the brake fluid level and its condition in both sections of the master cylinder or reservoir. The fluid level will drop as brake pads wear, resulting in a lowering of the brake fluid in the reservoir. Visually inspect the entire brake system for leaks. The reservoir on most vehicles is of the see-through design so the cap or lid does not have to be removed for checking. Most brake fluid types are hygroscopic, which means they absorb water/humidity. Also, over time, brake fluid picks up metal and other foreign matter, so it may need to be replaced periodically.

Notice on top of this brake master cylinder the see-through reservoir which enables the brake fluid level to be checked without exposing the brake fluid to contaminants. Note also the brake warning light wiring and brake line connections which should be inspected.

Check the condition of the brake fluid by looking at it or by testing its condition. Most brake fluids appear clear; however, some may be amber or red colored. If the brake fluid is murky brown or contains impurities, flush the hydraulic braking system and replace the fluid. Test the fluid properties with a brake fluid test strip or by checking its electrical (galvanic reaction) properties using a digital voltmeter. Generally, there should be no more than 0.3 volts between the fluid itself and the metal master cylinder housing (electrical ground).

3. Inspect flexible brake hoses, brake lines, valves, and fittings for routing, leaks, dents, kinks, rust, cracks, or wear; inspect for loose fittings and supports; determine needed repairs.

Flexible brake lines should be checked for areas of brake fluid seepage as indicated by darker, moist areas. Vibration under the hood can stress even steel brake lines. Check for external leaks along the lines and around the brake line fittings. Tighten the fittings to specifications using a flare wrench. Do not over tighten a fitting in an attempt to stop a leak; rather, a new fitting, or the entire brake line (complete with new fittings) should be installed. Any brake line dents, kinks, rust, or cracking is not permitted and is cause for brake line replacement.

4. Verify operation of brake warning light and ABS warning light; inspect brake system wiring damage and routing.

Consult the vehicle owner's manual or service manual for the applicable on-board brake warning indicators, and with the ignition turned on, note if all the brake warning lights illuminate as they should during the self-test sequence. The brake warning light will be red and the ABS warning light will be yellow. A red light may indicate low fluid, a sticking sensor float in the brake fluid reservoir, or simply that the parking brake is set.

This image shows an IPC during a KOEO self-test sequence. Getting past all the confusion, note the top center red brake warning light and the yellow ABS light.

While under the hood, make certain the brake warning system wiring and its connections on the master cylinder are intact and damage free. Check that the brake warning circuit wiring is not near the hot exhaust manifold, the EGR plumbing, or other areas of concern.

5. Test parking brake indicator light, switch, and wiring.

While behind the wheel, apply the parking brake, making certain it activates the brake warning light or dedicated indicator lamp. Make certain its wiring is properly secured behind the dash or under the center console. If the brake light remains off when the parking brake is applied, check the fuse. Disconnect the wiring from the brake light switch and ground it; the light should come on. If it does, check the switch with a DVOM for continuity. If the light stays on all the time, also check the switch. Check its adjustment by cycling the parking brake several times to see if the light goes OFF. If it still remains on, disconnect all switches that activate it one at a time: park brake switch, low-fluid switch, pressure differential switch, and the ignition.

6. Bleed and/or flush hydraulic system.

Following the vehicle maintenance schedule, or as needed, perform a brake system flush and refill according to the manufactures recommended procedure. Brake system flushing may be done by using a special brake flush machine or by using a pressure bleeder, vacuum bleeding, or manual bleeding. Follow the instructions for the pressure bleeder, which will force fresh brake fluid into the brake system as you open and close each wheel cylinder or caliper bleed screw one at a time until the brake fluid comes out clean.

7. Select, handle, store, and install proper brake fluids.

There are various types of brake fluid, each designed for specific vehicles and their brake system needs. Be sure to use only the fluid the OEM recommends for a given vehicle. Many vehicles use DOT3 fluid, which is hygroscopic and generally clear in color. A few vehicles may require silicone-based fluid or another brake fluid. Brake fluid kept in a sealed container does not generally go bad, but once the container or can has been opened, the fluid can quickly absorb water from the ambient air. Hence, when brake fluid is stored, the containers should be kept tightly sealed. Brake fluid can ruin the paint of a vehicle, so avoid spills on the vehicle when adding to or refilling the brake system. Pour fresh brake fluid into the master cylinder reservoir using a clean plastic funnel, and only until the fill line is reached. Then, promptly close the master cylinder reservoir.

Drum Brakes

8. Remove, clean, inspect, and measure brake drums; follow manufacturers' recommendations in determining need to machine or replace.

Replacing brake shoes and hardware can be a dirty and difficult job until the service technician gains some experience. Brake drums may or may not include the wheel hub. Follow the manufacturer's instructions for how to remove them. When removing and cleaning brake drums, use personal protection equipment (PPE) and OSHA-approved devices and methods that capture the hazardous brake dust and minimize risk to personnel.

Once cleaned, check the drums for hard spots, hot spots, concave or bell mouth surfaces, or deep grooves or scoring. If cracked or damaged, discard the drum and replace it. Determine if the drums will require their friction surfaces to be cleaned up on a brake lathe. Use a brake drum micrometer to determine if the drum is warped or out of round, or if the maximum wear limit (discard dimension) has been reached or exceeded (or would be, once the drums are "turned"). Adhere to the drum manufacturer's wear limit specifications that are found stamped or cast on the drum.

9. Machine drums according to manufacturers' procedures and specifications.

Follow the brake lathe manufacturer's instructions when resurfacing brake drums. If it is not a single cut machine, a finish cut may be needed. Avoid removing too much material in one pass or cutting too fast, which creates a "threaded drum" and causes the brake shoes to follow the cut and make a "snapping noise" as the brakes are released.

10. Using proper safety procedures, remove, clean, and inspect brake shoes/linings, springs, pins, self-adjusters, levers, clips, brake backing (support) plates, and other related brake hardware; determine needed repairs.

Follow OSHA regulations and approved methods of brake dust control. Wear PPE (goggles, gloves, and paper mask) when servicing wheel brakes. After removing all drum brake hardware and shoes, determine if major components (self-adjuster linkage, adjuster screw assembly, and wheel cylinder) need replacing. The springs and hold-down parts are generally supplied in a kit and replaced when rebuilding/replacing drum brakes (shoes).

If reusing the springs, inspect for warped or bent coils, nicks, and a bent or twisted shank. Use soap and water and a wire brush to clean the backing plate and any attached parts. Catch all runoff in a pan and dispose of it properly.

11. Lubricate brake shoe support pads on backing (support) plate, self-adjuster mechanisms, and other brake hardware.

Once the backing plate is clean and dry, lubricate the pads where the shoes rest with specially formulated water-resistant grease. Avoid getting grease on the drum friction surface or the brake shoes' lining surfaces. After cleaning the threads on a wire wheel, use the same grease on the adjuster screw, and apply grease to other friction surfaces of the parking brake and self-adjuster mechanisms.

12. Inspect wheel cylinder(s) for leakage, operation, and mounting; remove and replace wheel cylinder(s).

Check the wheel cylinders by prying back gently on their boots. Check for signs of brake fluid inside, which would indicate the cups are leaking. Wheel cylinders may be rebuilt, but it's difficult to find rebuild kits. For this reason, the entire wheel cylinder is generally replaced. When replacing the wheel cylinder, use a flare nut wrench with caution when loosening and removing the brake line fitting from the wheel cylinder. It's easy to twist the fitting off the brake line! If this happens, a section of the brake line must be replaced. Clean all mounting surfaces and fasteners before installing a new wheel cylinder assembly.

13. Install brake shoes and related hardware.

With everything clean and inspected, have a replacement brake parts kit on hand, along with the proper lubricant. Have the tools needed (brake spring pliers, hold-down spring tool, brake spoon, and pre adjustment gauge) on hand. Having the right tools makes "hanging brakes" a lot easier.

Keep your hands grease free when installing the brake shoes themselves. Make certain not to reverse the placement of the primary and secondary shoes of duo-servo brakes systems. The brake shoe with the longer lining (the secondary shoe) goes towards the rear of the vehicle.

14. Adjust brake shoes and parking brake.

Once the shoes and all hardware are installed, preadjust the shoes with the star wheel adjuster according to the drum diameter as determined by using the brake shoe preadjustment/set gauge.

With everything installed, determine if the drum brakes need adjusting by pumping the brake pedal twice with the engine running. If the pedal rises (less free play) on the second pump, the drum brakes need to be adjusted.

Follow the manufacturer's instructions for making the final drum brake adjustment.

At the backing plate access hole, use a brake adjusting tool ("brake spoon") on the star wheel adjuster to tighten and center the shoes. On self-adjusting brakes, raise the adjuster lever with a small flat screwdriver while rotating the star wheel. Periodically bang on the drum to assist the shoes to seat. Tighten the adjuster star wheel until the drum (or tire and wheel assembly) can no longer be rotated. Then, back off the adjuster star wheel until the tire assembly (or drum) turns freely. Be sure to replace the rubber access hole plug.

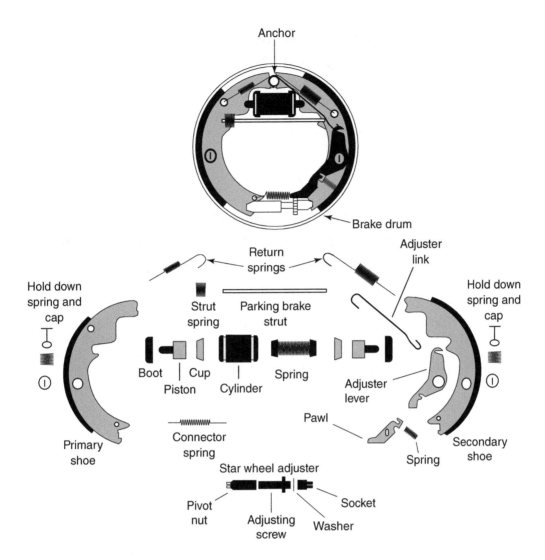

15. Check parking brake system operation; inspect cables and components for wear, rust, and corrosion; clean or replace components as necessary; lubricate and adjust assembly.

Apply and release the parking brake, making sure the wheels rotate freely when released. Set the parking brake fully with the vehicle on a mild uphill and a mild downhill (both < 30 degrees) and determine if the brake holds the vehicle stationery. If the brake does not hold the vehicle, adjust the parking brake linkage according to the vehicle manufacturer's instructions. Some parking brakes are adjusted at the rear backing plate, others at the cable "equalizer" beneath the vehicle.

Inspect all linkage and cables for rust, kinks, or damage that could bind the parking brake cable in its conduit and prevent it from releasing. Check the parking brake mechanism and use an approved lubricant on it. Check the cable for broken strands, rust, or kinking. Replace it if it is damaged in any way.

Adjust the parking brakes only after the service brakes have been adjusted. Follow the OEM procedures, which generally involve adjusting the parking brake equalizer under the car. Following any adjustments, test the parking brake as described above.

16. Reinstall wheel, torque lug nuts, and make final brake checks and adjustments.

Install wheel fasteners by starting/threading them by hand, then lightly wrench-tighten them. With the vehicle on the ground, torque the fasteners to factory specifications in an alternating sequence (star or criss-cross). Check the brake pedal for proper height and firmness. If the pedal is too low, recheck the drum brake adjustment, making sure they do not drag. Also refer to Task D.52

Disc Brakes

17. Retract integral parking brake caliper piston(s) according to manufacturers' recommendations.

Some vehicles with 4-wheel disc brakes use a miniaturized version of a drum brake system as parking brakes. The rear rotors serve as drums for the brake shoes, and adjustments are done with a star wheel adjuster.

The more common caliper-actuated parking brake mechanically applies and locks the brake pads against the rotors. While a variety of methods are used, all are lever actuated from the inboard side of the caliper. To retract the brake pad away from the rotor so brake service may be performed, the caliper piston must be screwed back into its bore in the caliper by using a special spanner-like tool. Depending on the vehicle application, follow the manufacturer's instructions and use the recommended tools.

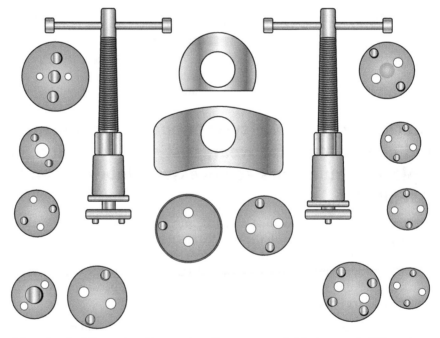

All sorts of piston retracting tool options are available to meet different vehicle needs.

18. Remove caliper assembly from mountings; inspect for leaks and damage to caliper housing.

The brake caliper must be loosened and most likely removed from its mountings in order to service the brake pads, the rotors, or the caliper itself. The specific procedure for the vehicle being serviced should be located in the OEM service instructions and followed carefully. Generally speaking, remove some brake fluid from the reservoir first. Some mechanics advocate opening the caliper bleed screw slightly to avoid pushing brake fluid back up the brake lines. Use a C-clamp or special tool to force the caliper piston(s) fully back into the bore.

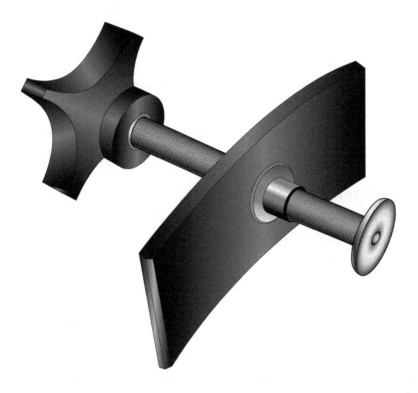

Use a piston retracting tool like this, or a C-Clamp, to retract the piston into its bore so the caliper can be slid off the rotor and be removed.

Disconnect the brake hose, removing the mounting brackets and, on some calipers, sliding bolts, or pins. Pry the caliper loose from the rotor and lift it from the rotor assembly. Do not let it drop.

Inspect the caliper for leaks at the piston dust boot. If possible, peel back the boot and inspect for a leaking piston seal. If used, check the pin boots and bellows for tears. Check for physical damage from impact with road debris or from a collision.

19. Clean and inspect caliper mountings and slides/pins for wear and damage.

Sliding caliper slides and pins should be cleaned with brake parts cleaner and inspected for physical damage, nicks, burrs, or other irregularities. Check for rust, corrosion, pitting, or other problems. If any of these parts are damaged, pitted, corroded or rusted, replace them.

20. Remove, clean, and inspect pads and retaining hardware; determine needed repairs, adjustments, and replacements.

On some vehicles, the brake pads can be released and slid out of the calipers without loosening or removing the brake caliper. On most vehicles, the caliper will need to be dismounted from the spindle and hung up to avoid brake line damage. With the pistons retracted, the pads can then be removed from the caliper. On floating calipers, pins, keys, or other devices that are securing the pads need to be removed. Follow the OEM service procedures for removal of the pads from the caliper, and take note of their location. If the pads are to be reused, check the condition of the brake pad linings and the flanges. Generally, the pad should have a minimum lining thickness of 0.125 inch (1/8 inch or 3.175 mm) or more on the backing plate if it is to be reused.

> **Note:** *Repair one wheel at a time so you can refer to the other side of the vehicle if unsure about how things should go back together. Some replacement parts are typically supplied with new replacement brake pads; be sure to use them instead of the old ones.*

Clean all parts with brake cleaner and inspect all parts for rust or corrosion. Replace any that show wear or damage. If one pad has worn much more than the other on a given wheel, inspect the sliding caliper for freedom of movement and re-lubricate or replace parts as needed.

21. Clean caliper assembly; inspect external parts for wear, rust, scoring, and damage; replace any damaged or worn parts; determine the need to repair or replace caliper assembly.

Use brake cleaning solvent and a wire brush to remove road debris and rust from the caliper assembly. Be sure to wear goggles when doing so. Check all parts of the caliper for damage. Pay special attention to the mounting ears. If they are damaged or cracked, replace the caliper. Check the pins and slide areas for nicks or other irregularities that might prevent parts from moving freely. Check all areas for rust or corrosion and wire brush them as required.

22. Clean, inspect, and measure rotors with a dial indicator and a micrometer; determine the need to index, machine, or replace the rotor.

Clean the rotors with brake cleaning solvent and inspect them for hard spots (small blue areas) or for heat checking (small spider-like cracks) caused by the rotor's getting too hot. Check for deep grooves, scoring, cracks, or other damage. If deeply grooved (more than 0.015 inch), resurface the rotor on a brake lathe or replace it if doing so will reduce its thickness close to its discard thickness dimension (within ~.030 inch). Look for the discard thickness dimension cast or stamped on the rotor in thousandths of an inch or in millimeters. If minor rust is evident, remove it with emery cloth. If the rotor is heavily rusted, consider turning it on a brake lathe. If the rotor is cracked, scrap it. If so equipped, check the built-in wheel speed sensor carefully for damaged or missing teeth.

With the rotor secured by the wheel fasteners to its hub on the vehicle or secured in place on a brake lathe, check for rotor runout/warpage with a dial indicator. Check your measurement against the OEM specifications. Using OEM procedures, employ an inside micrometer to measure for uniform rotor thickness and parallelism. Compare your readings to OEM specifications. If the rotor is out of specifications in any way, consider turning it on a brake lathe or discarding it. If the rotor does not meet the discard thickness dimension, it needs to be scrapped.

23. Remove and replace rotors.

First, remove the brake caliper, as noted in Task E.18. Before removing the rotors, mark them with "Left" or "Right" for correct reinstallation. Use penetrating oil to help separate a rusted rotor from its mating hub. Sometimes tapping on it (not on the studs) will help free it. Once removed, separate two-piece rotors (rotor and hub).

When reinstalling a rotor, make certain all mating surfaces are free of rust and scale. Use a wire brush if necessary to ensure a clean surface. Any dirt or rust particles not removed will potentially create run out problems later on. Keep the rotor friction surface clean and free of grease and other lubricants. Remove the protective coating using soap and warm water, and wipe the surface dry with a lint free cloth.

24. Machine rotors, using on-car or off-car method.

Machining the rotor on the car ensures near zero run out and generally saves the labor time of removing the rotor. Off-car brake lathe machining can also provide new life to an otherwise non-useable rotor.

Debate goes on whether or not the rotor should automatically be turned whenever it is removed from the vehicle. Turning rotors may be profitable, but removing any metal leaves the rotor thinner and less tolerant to recovering from the heat of braking (friction). On the other hand, leaving existing grooves and scoring on the rotor means the new pads will not seat as readily, and they will take longer to break in.

Whether on or off the car, follow the machine/brake lathe manufacturer's rotor machining instructions to the letter. When machining, remove as little metal as necessary to get a smooth rotor surface. Make certain the cutting bits are in perfect shape. Remove the same amount of metal from both sides of the rotor and from both same-axle rotors if possible. Avoid making too-deep cuts or spiraling the cut by advancing the bits too fast.

When finished, wash down the newly machined rotors with soap and water to remove any metal filings from the rotor surface. Towel them dry.

Use an on-car brake lathe or one like this if the rotors have been removed from the vehicle.

25. Install pads, calipers, and related attaching hardware; lubricate components; bleed system.

Unless there is a good amount of life left on used pads and they are in good shape, install new brake pads on the vehicle. Make sure all brake pad hardware, fasteners, and pins are clean and free of defects.

> *Note:* So-called organic brake pads are inexpensive; metallic pads last longer, but they may be noisy; and ceramic pads cost more, but they reportedly better handle the heat encountered during long downhill descents or when towing. Newer and better braking compounds are becoming common.

Follow the manufacturer's instructions when replacing the brake pads in the rotors. Make certain the correct pad is installed on the outboard side of the vehicle. Insert the pads into position and assemble all attaching hardware, including spring clips and anti-rattle devices. Make sure the pads are firmly seated, and bend the tabs (if used) as required for a firm fit. Some technicians use noise quieting paste on the backing plates of the brake pads; if doing so, apply the paste sparingly and only where the backing plate touches other parts. Lightly grease the slides, if used, with approved grease or anti-seize compound so the pads can move inward as they wear.

Slide the caliper into position on its mounting and over the rotor. Install a mounting bolt and swing the caliper fully into position on the rotor. Install the other mounting bolt and torque both of them to specifications.

Bleed the brakes using either the pressure bleeding or the vacuum method. When pressure bleeding, be sure to hold open the metering valve or use a special tool to hold open the combination metering/proportioning valve. Bleed the brake system according to OEM procedures, which vary widely depending on the make, model, and year of the vehicle. Generally, the wheel farthest from the master cylinder is to be bled first, but not always.

Alternatively, when brake bleeding manually with an associate pumping the brake pedal, make certain the master cylinder has enough fluid at all times. Once the bleeding is completed, top off the master cylinder, clean its cover, and replace it.

26. Adjust calipers with integral parking brakes.

Follow the vehicle manufacturer's instructions for adjusting integral parking brakes. There may be an adjustment nut or cable adjustment at or near each caliper, or the parking brake may be self-adjusting (Ford). For calipers with integrated drum-type parking brakes, adjust the star wheel.

27. Fill master cylinder with recommended fluid; reset system; inspect caliper for leaks.

Make certain the brake fluid being used is fresh and of the required type. Using a clean plastic funnel, fill the reservoir to the full line and install the freshly cleaned cover. Wipe up any spills immediately. Use a factory approved scan tool to reset the electronic brake warning system once any faults have been properly repaired. Apply the brakes firmly numerous times and inspect for any leaks at the brake line connection to the caliper, at the bleed screw, and around the caliper piston and boot area.

28. Reinstall wheel, torque lug nuts, and make final brake checks and adjustments.

Install wheel fasteners by starting/threading them by hand, then lightly wrench tighten them. With the vehicle on the ground, torque the fasteners to factory specifications in an alternating sequence (star or criss-cross). Doing otherwise could cause rotor warpage and subsequent pedal pulsation.

Check the brake pedal for proper height and firmness. If the pedal is too low, pump the brakes a few times to seat the pads on the rotors and recheck the brake fluid level in the reservoir.

29. Road test vehicle and burnish/break-in pads according to manufacturers' recommendations.

Take the vehicle for a test drive, making sure all previously reported symptoms have been taken care of. Breaking- in (aka bedding-in or burnishing) the new brakes is an important step not to be overlooked. Burnishing transfers pad material evenly to the rotors and cooks off pad residue, resulting in long and quiet brake life. It also helps eliminate brake squeal.

Perform the burnishing procedure according to the brake material provider's instructions. Some call for repeated light stops; others call for repeated moderate-to-heavy brake applications with specified wait times in between. Up to 30 repeated stops from a specified speed may be called for to complete the process. Recheck the wheel fastener torque when finished with the test drive/break-in procedure.

Power Assist Units

30. Test brake pedal free travel with and without engine running to check power booster operation.

To perform the power booster performance test on a vacuum-assisted brake system, pump the brakes numerous times with the engine OFF to remove any residual vacuum from the power booster. Make note of the pedal position. With your foot still on the brake pedal, start the engine and note if the pedal sinks. If the brake booster is working properly, you should feel the pedal sink an inch or more. If the pedal does not change its position, the booster is not working properly and you will need to check the vacuum supply to the booster.

31. Check vacuum supply (manifold or auxiliary pump) to vacuum-type power booster.

Remove the vacuum line from the brake booster and attach a vacuum gauge to the line coming from the intake manifold. Start the engine and read the vacuum gauge. The gauge reading should match the intake manifold vacuum measured elsewhere; that is, greater than 16 inches of vacuum. If not, check for a leaking or pinched vacuum line to the brake booster, or for a clogged or stuck-closed in-line vacuum check valve.

32. Inspect the vacuum-type power booster unit for operation, and vacuum leaks; inspect the check valve for proper operation.

See if the power brake booster vacuum line is firmly attached to the brake booster. If an in-line check valve is used, make certain you can pull a vacuum with a hand vacuum pump in one direction only. If it is stuck open or closed in both directions, replace it. Alternatively, idle the engine with your foot on the brake pedal, then shut off the engine, keeping your foot on the brake. Keep your foot on the brake pedal for up to 10 minutes. The pedal should not rise. If it does, it indicates a vacuum leak.

Check the booster itself for rust or holes. Run the engine and then turn it off. Pull the vacuum line off the booster and listen for air, which should rush into the brake booster. With the engine idling, applying the brakes should not affect engine idle speed or smoothness. If the engine idling becomes rough or the engine stalls, check for a failed brake booster.

33. Identify operation of electric-hydraulic assist system; check system for leaks and operation.

On some import vehicles, an electro-hydraulic braking system is used. This system uses a high-pressure reservoir which supplies the required braking pressure quickly and precisely to the service wheel brakes even without the driver's involvement. Electro-hydraulic braking systems offer improved active safety especially when braking in a corner or on a slippery surface.

The so-called Sensotronic Brake Control (SBC) electro-hydraulic brake system was developed by Daimler and Bosch. With electro-hydraulic braking, the driver doesn't directly activate the brakes. Rather, the master cylinder is activated by an electric motor or pump that is regulated by an electronic control unit. When the driver hits the brakes of an electro-hydraulic system, the control unit processes information from a number of sensors to decide how much braking force each wheel actually needs. The system then applies the necessary amount of hydraulic pressure to each brake caliper.

The system also includes functions to reduce the driver's workload, including Traffic Jam Assist (which brakes the vehicle automatically in stop-and-go traffic once the driver takes his or her foot off the accelerator) and a Soft-Stop function (which allows particularly soft and smooth stopping in city traffic).

Electro-hydraulic brake systems typically operate under much higher pressures than traditional systems. Where hydraulic brakes normally operate at around 800 PSI, the Sensotronic system operates at between 2,000 and 2,300 PSI.

34. Identify operation of hydro-boost assist system; check system for leaks and operation.

Check the fluid level and add fluid if needed. If the fluid is low, check for leaks. To test the hydro-boost system, with the engine OFF, pump the brake pedal 10 or more times to remove residual hydraulic pressure from the system's accumulator. Push firmly on the brake pedal and start the engine. The brake pedal should sink at first, then it should then push back against your foot. If it fails to do so, the system is not providing hydraulic boost.

To check the hydraulic accumulator, start the engine and turn the steering wheel against its lock. Hold it there for about 5 seconds, then release it. Shut off the engine, and wait for 30 minutes or more. Press on the brake pedal 2 or 3 times to check for power assist to the brakes (easy brake application). If power assist is not available for at least 2-3 applications, the accumulator is not holding pressure. Check for leaks along all of the hydraulic lines to/from the accumulator, as well as their fittings.

F. Electrical (8 questions)

1. Disarm/re-enable air bag; verify lamp operation.

Service to the air bag system requires disarming the air bag deployment circuits. Otherwise, accidental air bag deployment could occur, causing serious personal injury. Check the OEM service manual and follow its procedures to the letter before performing any work on the Supplemental Restraint Systems (SRS) to avoid accidental deployment of the system. Technicians are too often injured by accidental air bag deployment.

In general, disabling the front and side/side curtain air bags requires turning off the ignition, removing the SRS fuse(s), disconnecting various connectors (yellow colored) behind the dash and under the seats, and waiting the required time until the system's module powers down. See also Task D.1.

To re-enable (rearm) the air bag system, follow the OEM instructions. Reconnect any disconnected connectors, and replace the SRS fuse(s). With everything reconnected, turn on the ignition and make certain the yellow SRS light on the IPC self-tests—then goes out. If it does not, use the appropriate scan tool to pull any SRS trouble codes which exist and make note of them for follow-up service by a qualified technician.

2. Check voltages, grounds, and voltage drops in electrical circuits; interpret readings.

Think of a voltmeter as a balance-type scale, comparing electrical pressure from two places.

Voltage drop is the difference in electrical pressure (think again of the scale) between two points within an electrical circuit. To check a system or circuit for available voltage (electrical pressure, or EMF) dynamic resistance, connect the DVOM's leads across the load or cable to be tested, power up the circuit, and note the reading. For example, to test for resistance in a starter motor cable, connect the positive lead to the positive battery post and connect the DVOM negative lead to the starter solenoid's B+ terminal. While cranking the engine, note the voltage drop (lost electrical pressure) on the meter.

Check your readings taken against the OEM specifications.

> *Note: In accord with ohm's law, given a fixed amount of resistance, the greater the amount of amperage that is flowing in the circuit, the greater the voltage drop will be. This is why taking dynamic voltage drop readings is a superior method for finding unwanted resistance than using an ohmmeter to find static resistance.*

3. Check current flow in electrical circuits and components; interpret readings.

To measure amperage, the circuit must be live (current/amps flowing). Connect the appropriate inductive amps clamp (high or low amperage) around the cable or wire to be monitored for current flow. Be sure to observe the polarity markings on the clamp; if placed backwards on the wire or cable, the polarity read on your meter will be backwards. Power up the circuit and take your reading, scaling down as required. Compare your readings to specs.

If a current clamp is not available, you will need to "break into" the circuit to read amperage flow; current/amperage is always measured in series when measured from within the circuit. Be sure to choose the correct "DC Amps" meter lead cavities on your DVOM.

A DVOM will typically only read up to 10 amps (or less, check your meter), so make certain your meter is appropriate for what you are trying to measure; for example, you would NEVER attempt to read starter current amps (as high as 200 amps or more) using a meter wired in series.

Insert the DVOM leads at the place you broke the connection—at a connector or component—observe the correct polarity. The positive lead should be on the source-of-current side of the circuit; that is, closer to the battery.

4. Check continuity and resistances in electrical circuits and components; interpret readings.

Use the Ohms scale of the DVOM for taking resistance measurements. Make certain the circuit has no power in it, because the ohmmeter supplies its own power from its internal battery. A live powered-up circuit will distort the meter's resistance reading, or even destroy the meter! You will likely have to isolate the load or remove it entirely from the circuit to get an accurate resistance reading. Resistance readings are always taken in parallel across the load (resistance) being measured.

5. Perform battery tests (load and capacitance); determine needed service.

> **Note:** *A two-minute period should be maintained between any load testing cycles to avoid damage to the battery.*

The battery load test is an effective and dynamic way to determine the capability of a battery. The test demands that the battery voltage deliver a specified amount of current (in amps) over a given amount of time without falling below a specified voltage potential. Thus, the load test determines if a battery can deliver when the going gets tough. It is done by placing a heavy load across the battery terminals. The load itself could be an actual starter motor cranking the engine, the vehicle's headlights, or an external load such as a carbon pile built into a volt-amp tester (VAT). In much earlier days, a simple load tester consisted of nothing more than a heavy duty "heater grid" with an analog voltmeter. The heavy resistive load was placed across each individual battery cell to essentially "short it out" and test its ability to sustain its voltage.

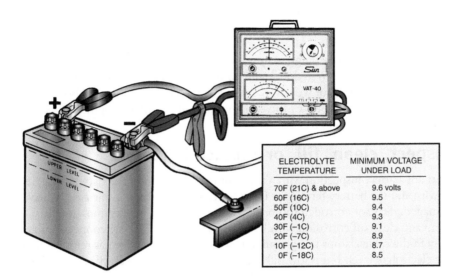

ELECTROLYTE TEMPERATURE	MINIMUM VOLTAGE UNDER LOAD
70F (21C) & above	9.6 volts
60F (16C)	9.5
50F (10C)	9.4
40F (4C)	9.3
30F (−1C)	9.1
20F (−7C)	8.9
10F (−12C)	8.7
0F (−18C)	8.5

It is important to start a load test with a fully charged battery. Typically, a load equaling one half of the battery's cold-cranking amperage CCA rating is placed on the battery for 15 seconds. The battery should still be above 9.6 volts at the conclusion of the test. If not, recharge and retest the battery. If it fails again, recycle the old battery and replace it.

Modern electronics has made battery testing faster, easier (and arguably), as reliable as load testing with the use of the capacitive-conductive battery tester. After entering pertinent battery data into the tester, the tester measures the battery's plate conductance (which represents the battery's internal condition), and issues a report of Pass, Retest, or Fail. Unlike with a load test, the capacitive-conductive tester can even check the battery plates' condition when the battery is low on charge. Capacitive-conductive testers are said to be able to tell the useful life remaining in the battery. Be sure to read and follow the tester manufacturer's instructions to ensure a reliable battery test.

> *Note: When testing AGM batteries, take note of any special requirements when entering the battery's CCA rating.*

6. Maintain or restore electronic memory functions.

If battery service is required, removal of the battery connections will probably erase any keep alive memory (KAM) data stored in the various on-board electronic modules. The engine and transmission may perform erratically when first started after battery power has been lost and restored, these must undergo a relearning process as the vehicle is once again driven. Likewise, with a loss of KAM, DTCs stored in the PCM are erased and the OBD-II monitors must rerun before a state-mandated emissions test can be performed.

Once KAM is lost, it cannot be retrieved. All lost settings must be reestablished, either through normal vehicle operation or by manually resetting (ex: radio presets), and the vehicle owner is inconvenienced at best. Some on-board devices with theft-deterrent features will need to be reactivated using a special code. Refer to the vehicle owner's manual or service manual for specific instructions on how to perform a reset procedure and how to obtain the reset code(s) needed.

What all this means is that it is easier to avoid erasing KAM in the first place. The secret is to provide the vehicle with constant auxiliary 12-volt power before disconnecting the service battery. Auxiliary power can be provided via the data link connector (DLC) on 1996 and newer OBD-II vehicles, or via the assembly line data link (ALDL) used on pre-1996 OBD-I vehicles. Alternatively, on vehicles with B+ powered power outlets, a 12-VDC adaptor can be plugged into the power outlet/cigar lighter socket to maintain KAM. Remember that in so doing, the positive battery connection will be live with power, so keep the B+ connector well away from ground while cleaning it. Also, avoid opening the doors or activating the key fob that will turn on the interior lights and possibly overtax the 12-VDC power supply.

7. Inspect, clean, fill, or replace battery.

Start any battery inspection with a thorough examination of the battery case, its hold-downs, and the battery terminals. Look for cracks or leakage in the case. Check for damaged, dirty, or corroded top or side terminals. If the connections are dirty, wire brush them and clean off corrosion with baking soda and water. If the connections are clean, tighten the cable clamps using a suitable wrench. If the battery has removable caps, check the electrolyte level. If it is low, fill each cell to the proper level with distilled water. Be careful to not let impurities fall into the battery cells.

When replacing a battery, be sure to follow standard safety procedures. Remove the negative cable connection from the battery first (on a negative-ground vehicle) to avoid a short circuit caused by the wrench's touching against a metal ground. Batteries are rated by group number or their physical size, + and − post placement, and ampere/hour capacity. Make certain the battery being installed is the correct size and rating intended for the vehicle.

8. Perform slow/fast battery charge in accordance with manufacturers' recommendations.

If fully or heavily discharged, 12-volt vehicle batteries are best brought back up to a full charge by using an external battery charger. Do not rely on the vehicle's charging system to do so; the strain on the vehicle's alternator to bring a dead battery up to full charge could weaken and, sooner or later, cause the diodes in the alternator to fail.

Different battery types require different charging algorithms. Normal flooded lead-acid batteries can be slow charged, or fast charged within limits to bring them up to full service voltage (12.6 volts). A slow overnight charge is always best and easiest on a battery. If rapid charging must be done, be sure to use a voltage and amperage regulated charger, and set to the proper charge feature for the respective battery. Valve Regulated Lead-Acid (VRLA) may be of the gel type or Absorbent Glass Mat (AGM) type battery. These require a different charging algorithm (charging cycle), so check with the vehicle manufacturer or the battery maker for instructions on the safest, most efficient, and reliable charging method for these types of batteries.

9. Inspect, clean, and repair or replace battery cables, connectors, clamps, and hold-downs.

Automotive service battery connections should be periodically removed and cleaned of any buildup of residue or corrosion. Use a special battery cleaning wire brush on battery posts, side connections, and cable ends. Use baking soda and water to neutralize and wash away remaining caustic residue (green and white contamination). Follow standard safety and KAM retention practices (see Task F.7) when removing and installing the battery connections.

Inspect the battery cable ends for deterioration from corrosion or rust damage, and inspect the clamps or bolts for damage from abuse or over-tightening. Inspect the cables upstream from the cable ends for hidden corrosion under the insulation. Cut off and replace (by soldering with a torch) defective cable ends, following the supplier's instructions carefully. Only in emergency situations should bolt-on replacement cable ends be used; they are prone to short-term failure.

A 12-volt flooded lead-acid battery's life is shortened if the battery is free to bounce around under the hood (AGM and spiral cell batteries can take more of a beating). Still, be sure to check the battery hold-down devices for rust and corrosion. Wire brush and paint them as necessary or replace them altogether. Check that the elongated hold-down bolts grip as they should, and that they are not corroded or stripped. Replace them if needed. It is a good idea to clean the battery tray as long as the hold-downs are being serviced.

10. Jumpstart a vehicle with a booster battery or auxiliary power supply.

If a vehicle battery has been run down or seriously discharged from a parasitic load such as leaving the lights on, the vehicle will either require a fully charged replacement battery or a jump-start. Jump-starting can be done from an auxiliary power supply that provides a safe means of boosting a low or dead battery. A portable power supply comes complete with clamp-on battery cables, an ON/OFF switch, and reverse polarity protection. An alternative is to use standard jumper cables from a host vehicle with a good battery; it is not a good practice to rely on a battery charger for jump-starting.

Always connect the negative cable last to a good ground under the hood of the vehicle with the dead battery. Avoid connecting the cable near the battery, which may have vented explosive hydrogen gas when being discharged. Turn on the switch of the power supply and let the power supply transfer power to the dead battery for a couple of minutes. If using a host vehicle, run the engine at a fast idle for several minutes to partially restore a charge to the dead battery. Then, crank and start the dead battery vehicle. Shut off the power supply, first remove the negative (ground) cable previously clamped to ground, and then move it away from the dead battery. When removing the cable, do not temporarily clamp one cable to the other. The jaws of one clamp can bite through the insulation of the other cable, resulting in a massive short and undesirable results.

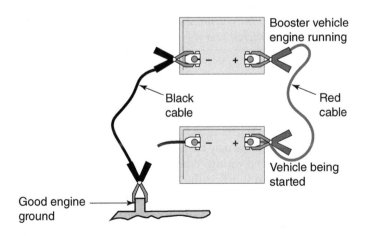

11. Perform starter current draw test; interpret readings.

> *Note:* *A two-minute period should be maintained between starter draw testing cycles to avoid damage to the battery.*

Starter current draw can be measured using a high-amp clamp on the negative battery cable or an inductive-type amp meter placed on the cable. Install the clamp so its arrow points away from the battery negative terminal. Use a remote starter switch or disable the vehicle's fuel supply by pulling the fuel pump relay. With the battery fully charged, crank the engine and read the amperage on your meter. Check your reading against the OEM specifications. In general, a V-8 gasoline engine starter will draw >200 amps, a V-6 will draw around 160 amps, and a 4-cylinder engine will draw around 120 amps. A diesel engine will draw considerably more amperage due to its higher compression ratio. Check the vehicle service manual for cranking motor test specifications.

An ampere reading lower than normal current draw could mean a discharged battery, excessive resistance in the battery and starter high-amperage cables, inside the starter or its solenoid, or in the cable connections. Crank the engine and check for voltage drops. Higher than normal amperage draw could be caused by an internal short in the starter motor or by excessive drag caused by failed engine bearings or other mechanical faults. A newly rebuilt engine will draw a bit more until it is broken in.

12. Inspect switches, connectors, and wires of starter control circuits.

Give special attention to the wiring and connections of the starter control (low-amperage) circuit. Neutral safety switches and starter solenoids under the car are subject to dripping oil, road splash, and other types of abuse.

Check all connections for tightness. If the system is acting up, perform voltage drop tests along the circuit, from the starter switch to the starter solenoid.

Theft-deterrent systems may include sensitive wiring within the steering column on certain makes of vehicles that is known to go bad. Such wiring faults can cause a no-crank condition, even if the starter and its wiring are fine.

> *Note:* *Troubleshooting theft-deterrent systems can be complicated, and the task goes beyond the scope of the MLR technician's on-the-job task list.*

13. Remove and replace starter.

Starter removal can be easy or difficult, depending on the make and model of the vehicle; therefore, make sure you know the starter is bad before attempting to remove and replace (R & R) it. Other components or parts may have to be removed just to gain access to starter mounting bolts/fasteners and wiring. When removing a starter, be sure to disconnect the battery first to avoid shorting the battery powered cranking circuit wiring. Be ready to support the starter once it has been unbolted; it may be heavy. Make certain any shims, if used, are saved for installation of the replacement starter.

14. Perform charging system output test and identify undercharge, no-charge, or overcharge condition.

The charging circuit can be tested by using a simple voltmeter to perform the 3-point charging circuit test:

First, check open circuit battery voltage with the engine off. Next, check battery voltage with the engine running at a fast idle and no accessories turned on. Lastly, with the engine at a fast idle, check battery voltage with all of the electrical accessories (rear defroster grid, blower on high, high-beam headlights, and fog lights) turned ON. Even with everything ON, charging system voltage read at the battery should be upwards of ~14.2 volts.

Compare the voltage readings you get against the OEM specifications. If the voltage readings are too high, suspect a faulty voltage regulator or voltage sensing wire/circuit. Make sure the alternator B+ connection is secure, not loose or burned. If voltage at the battery falls off when a load is placed on it, check the battery. If it is good, check the alternator and its wiring.

You can perform a dynamic test of the alternator and its wiring using a volt-amp tester (VAT) to place a load on the system to simulate vehicle loads. With the engine at a fast idle, watch system voltage as you dial in a load. When the system voltage starts dropping, check the amperage draw on the system. It should meet or exceed specifications. If it does not, check all charging circuit wiring for voltage drops. If good, suspect a bad alternator or regulator (or the regulator circuitry if it is externally regulated in the PCM).

15. Inspect, adjust, and replace generator (alternator) drive belts, pulleys, and tensioners.

Check the V-belts, or the serpentine belt, for aging, cracking, and stretching. Check belt tension using an appropriate test method or a tension gauge. See that the belt tensioner is operating as it should and is of the proper type (over-running). The same applies to any idler pulleys. If they are noisy or not operating correctly, replace them.

When tightening alternator mounting bolts, it may be necessary to use a heavy screwdriver to apply tension to the alternator while tightening its mountings and adjusting V-belt tension. When doing so, be extra careful to avoid breaking off the alternator mounting ears or causing damage to the alternator.

16. Remove, inspect, and replace generator (alternator).

Inspect the alternator for any physical damage. Perform a functional check of its operation. If it needs to be removed, follow the OEM instructions. Disconnect the battery, being careful not to drop it or damage the mounting ears or electrical connections. Some vehicles may require the half-axle or other components to be removed in order to remove and replace the alternator, so make sure your diagnosis is correct before calling for a replacement.

17. Inspect, replace, and aim headlights/bulbs and auxiliary lights (fog lights/driving lights).

Check for dim or burned out headlights, driving lights, and fog lights. If any are not working, first remove the non-operating bulb and check its continuity with an ohmmeter. If it is good, check the socket for voltage with the circuit or using a voltmeter. Check carefully for corrosion in the sockets. If no power is being provided, check the fuses and relays. If both lights are not working, check for a bad common ground connection.

If both headlights are dim, the battery could be low on charge or the battery connections could be dirty. If working on Xenon or high intensity discharge lamps, be extra cautious of high voltages that may be present. Follow the OEM service manual when checking these systems.

Use a headlight aiming machine for adjusting the headlights and auxiliary lights. Today's computerized machines make quick and easy work of headlight aiming–if the adjustment screws cooperate. Make certain the vehicle is on a level shop floor or driveway, that the tires are properly inflated, and that the suspension is properly aligned. The vehicle should be typically loaded and the gas tank about half full because all of these affect headlight aiming.

Some vehicles include spirit levels to assist in headlight aiming; some even have automatic headlight aiming and adjustment using servo motors. Use a scan tool to test these systems for proper operation.

18. Inspect interior and exterior lamps and sockets; repair as needed.

Check all interior lamps for correct operation. Check for proper BCM time-out functions of interior lighting. If any interior incandescent or LED lights are not working, check them. If they are burned out, incandescent bulbs can be replaced in the field. LEDs require a more in-depth repair; they must be resoldered, or a circuit board must be replaced.

Check all exterior lamps for correct operation. If an exterior light (including a headlight), is not working, first check for a burned-out bulb, then for a blown fuse. If equipped, check if a lamp out module indicates a malfunction. Check if an aftermarket trailer wiring harness has been added that may have upset power supplied to the tail or brake lights.

Check that full voltage is supplied to the light socket. If some voltage is present, the circuit is at least working so check for voltage drop on the circuit's hot side. Check for corrosion in the socket. If full voltage is present at the socket, check for a faulty ground. The problem also could be an open or a short in the wiring, or loose connections.

Check the brake lights, backup lamps, turn signals, and marker lights for proper operation. If not working, check the bulbs before starting to take things apart. Make certain the correct bulbs are installed; some look alike but have different pin configurations.

19. Inspect lenses; determine needed repairs.

If the headlight (or other) plastic lenses are cracked or broken, replace them. If the lenses look cloudy or fogged over, special kits and compounds can be used to polish them back to being nearly clear once more. If the lenses or light buckets have water collected on the inside, check for leaks where rain or snow is finding its way into the headlight (or taillight) assembly. Repair the leak, replace the gasket, or replace the light bucket assembly.

20. Verify instrument gauges, warning/indicator light operation; reset maintenance indicators.

Most warning systems rely on a grounding switch that turns the warning light on. When the switch closes, the warning light comes on. Watch the warning lights as they go through a self-test as soon as the ignition is turned on. If certain lights do not come on, they are either faulty, or the fuse to that circuit is blown.

The alternator warning light is not grounded at the instrument panel. Instead, current flows through the bulb and grounds inside the alternator. With the alternator charging, current flows through a wire to the opposite side of the bulb from the ignition switch. With both sides of the bulb receiving similar voltages, no current flows and the bulb does not light. If the alternator light glows slightly, check for a poor connection somewhere in the charging circuit.

Some maintenance indicators can be reset using a menu function on the driver's on-board information center. Others may be more involved, so be sure to follow the instructions in the vehicle service manual for specifics on resetting maintenance indicators.

21. Verify horn operation; determine needed repairs.

Operate the horn button and listen for a clear sound of the horn(s). If they fail to operate, check for a blown fuse and for faulty wiring (perhaps a rodent has chewed through the wiring). If the horn(s) is weak, check for resistance/corrosion, especially at the ground connections.

Modern vehicles activate the horn through the body control module (BCM), so using a scan tool to locate the fault may be helpful.

22. Verify wiper and washer operation; replace wiper motor, blades, and washer pump as needed.

Wet the windshield with the washer system and turn the wipers ON to make certain they operate smoothly and without streaking. Sometimes, merely cleaning debris from the rubber wiper blade insert is all that is needed to restore clear wiping action. If a wiper blade is cracked or hardened, replace the rubber insert or, better yet, the entire blade.

If the washer system fails to operate, listen for washer pump operation. If no sound is heard, check for a blown fuse or loose wiring. If the pump is working, the washer bottle may be out of fluid, or there may be a loose or cracked rubber feed hose leading to the spray heads. Check for clogged or mis-aimed spray heads; some can be cleaned and re-aimed using a straight pin.

G. Heating, Ventilation, and Air Conditioning (4 Questions)

1. Verify HVAC operation (vent temperature, blower and condenser fan, compressor engagement, blend and mode door(s) operation).

The first step in diagnosing a customer's HVAC complaint is to interview the customer and/or to carefully read the customer's concern as documented on the repair order. The next step is to put the HVAC system through its paces to make certain an actual malfunction exists. Sometimes a vehicle operator simply has not read or understood the vehicle owner's manual and therefore does not get the desired or anticipated results from the HVAC system.

The best way to check the system is to insert a thermometer probe into the center air conditioning-heating duct to make sure the system meets temperature delivery specifications. Listen for slow-to-fast blower speed operation as various selections are made. Listen for A/C clutch engagement as the A/C system or the defrost function is selected. Listen for belt noise or squealing as well. Also listen for the condenser or cooling fan(s) under the hood to turn on and off when A/C is selected or deselected.

In manual A/C systems, the temperature door is operated by a cable connected to the temperature lever, or by a vacuum motor.

In a semi-automatic A/C system, the temperature door is operated by an electric motor that is controlled by a module. The module receives inputs from the variable resistor connected to the temperature lever, the in-car sensor, and the outside air temperature sensor.

In an automatic A/C system, the temperature door, mode doors, blower speed, and compressor clutch are all operated by the climate control system's computer.

When putting the HVAC system through its paces, listen for the hissing of vacuum leaks or check for sticking or hard to operate Bowden cables.

2. Visually check A/C components for signs of leaks.

A/C systems can leak refrigerant over time either through tubing and their connections, or due to older, more porous, flexible A/C hoses. The telltale sign of refrigerant leakage is oil stains that appear at any of these locations, on the fins of the condenser, or elsewhere.

3. Inspect A/C condenser for restricted air flow.

The A/C condenser is typically located in front of the cooling system radiator. It is the first place leaves, bugs, and other debris can collect and block air flow through the condenser and the radiator. If any blockage exists, it must be cleared for both the A/C and the engine cooling system to work effectively.

4. Inspect and replace cabin air filter.

Many vehicles now include a cabin air filtration system to help keep incoming cabin air clean and free of debris (leaves, etc.) pollen and bacteria. Filters are not normally cleaned or reused beyond their recommended life. Maintenance schedules call for periodic replacement of cabin air filters, which are normally located under the passenger's side dashboard by the blower assembly. Some vehicles may require modifications to the dashboard facia in order to remove the filter for the first time. Alternatively, the filter may be located on the firewall under the hood or accessed outside from below the windshield wipers. Replacement involves sliding out the old paper filter and inserting a new replacement filter.

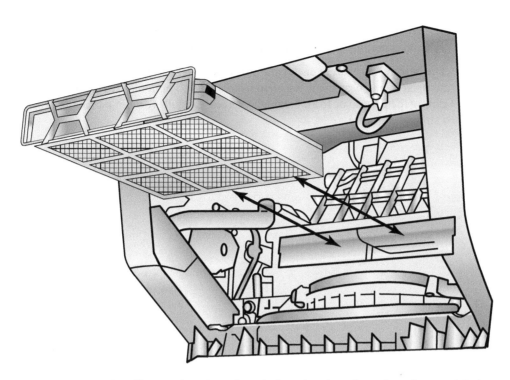

Replacing the cabin filter is often done from below the glove box. On other models they may accessed from the outside below the windshield.

5. Check drive belt for wear and tension; adjust or replace as needed.

While some more modern vehicles (including hybrids) use electrically driven A/C compressors, most A/C systems use engine-driven drive belts to rotate the A/C compressor. Older vehicles use V-belts; newer vehicles drive the compressor from the serpentine drive belt. Inspect these belts for wear, cracks, contamination, or looseness.

Gauges used for checking belt tension and serpentine belt wear are part of the MLR technician's toolbox. Serpentine belt manufacturers tell us that modern belt composition does not show the usual signs of dryness and cracking as they age. The proper way to test serpentine belts is by using a wear gauge. Belt tension is often done automatically by the tensioner, but older vehicles may require periodic tightening of the V-belt pulley to specifications. Make certain the serpentine belt has not jumped a groove on any of its pulleys and it is running in perfect alignment. If a complaint of a belt jumping from its pulley is heard, check for stuck, frozen, or misaligned pulleys along with proper tension.

When replacing belts, follow the routing guide typically found on a sticker under the vehicle's hood or in the service manual. Often when replacing a serpentine belt, it is a good idea to replace the tensioner at the same time.

6. Inspect and clean evaporator drains.

Condensate from the A/C evaporator must somehow exit the HVAC blower box below the dashboard and exit to the road. If leaves or dust have accumulated in the bottom of the box, the rubber tube leading to the outside may become clogged, causing a backup of water in the box. With the blower on HIGH, check for droplets of water spitting out of the heater vents or for emissions of foul smelling air. Also with the A/C on max setting look for water dripping on the front passengers floor.

In extreme cases, debris which has collected in the heater box may cause the evaporator to leak and cause the A/C system to lose refrigerant and fail.

The best way to prevent such failures is to keep leaves and other debris away from the cabin air inlets (near the windshield wipers) and to periodically clear the condensate tube exiting the firewall or from below the vehicle carefully (do not pierce the evaporator) with a stiff piece of wire.

Sample Preparation Exams

PREPARATION EXAM 1

1. While road testing a vehicle, a technician verifies the customer's complaint of steering wheel shimmy during hard braking. The technician should:

   A. tell the vehicle owner.

   B. tell the shop supervisor.

   C. tell the show manager.

   D. note the concern on the repair order.

2. While performing a scheduled maintenance test drive, which of the following noises would LEAST LIKELY be scheduled for repair?

   A. Belt squeal

   B. Brake squeal

   C. Tire noise

   D. Wind noise

3. Which type of replacement drive belt is shown in the illustration above?

   A. A "V" belt

   B. A serpentine belt

   C. A toothed timing belt

   D. A toothed "start/stop" drive belt

4. It's a good idea to routinely replace cooling system hoses:

   A. after 25,000 miles.
   B. after 50,000 miles.
   C. after 100,000 miles.
   D. based on the OEM schedule.

5. Engine coolant can be tested by using any of the following EXCEPT a(an):

   A. hydrometer.
   B. voltmeter.
   C. refractometer.
   D. ammeter.

6. Which of the following should illuminate if the charging system is not working?

   A. The ABS light
   B. The TPMS light
   C. The alternator light
   D. The check engine light

7. A vehicle's air filter is found to be contaminated with engine oil. Which of the following is the most likely cause?

   A. A leaking PCV hose
   B. Excessive blowby
   C. A stuck closed PCV valve
   D. A faulty idle-air control motor

8. When installing replacement spark plugs, some manufacturers recommend coating the threads with:

   A. chassis lube.
   B. motor oil.
   C. anti-seize.
   D. WD-40®

9. A pre-formed plastic pipe for the EVAP system is found to be cracked. Technician A says to wrap electrical tape around the crack to repair the leak. Technician B says to apply non-hardening sealant to the pipe. Who is correct?

   A. A only
   B. B only
   C. Both A and B
   D. Neither A nor B

10. An automatic transmission pan has been removed and is being reinstalled. Technician A says that some OEMs recommend using ATV sealant instead of a gasket. Technician B says that some OEMs recommend reusing the original gasket. Who is correct?

    A. A only
    B. B only
    C. Both A and B
    D. Neither A nor B

11. A CV joint boot is found to have a small split. Which of the following should be done?

    A. Seal it with ATV.
    B. Seal it with Permatex®.
    C. Replace the boot.
    D. Replace the half-shaft assembly.

12. A transmission cooler has developed a leak. Technician A says the cooler may be reliably repaired with epoxy. Technician B says the cooler may be reliably repaired with ATV sealant. Who is correct?

    A. A only
    B. B only
    C. Both A and B
    D. Neither A nor B

13. Technician A says that some automatic transmission filters are made of pleated paper. Technician B says that some transmission filters require removal of the transmission pan for replacement. Who is correct?

    A. A only
    B. B only
    C. Both A and B
    D. Neither A nor B

14. An oil-soaked engine mount:

    A. is OK and should be let alone.
    B. should be removed, cleaned and reinstalled.
    C. should be cleaned with spray solvent.
    D. should be replaced.

15. A manual RWD transmission is leaking at the rear seal. Technician A says the transmission will have to be removed from the vehicle. Technician B says that the drive shaft will have to be removed from the vehicle. Who is correct?

    A. A only
    B. B only
    C. Both A and B
    D. Neither A nor B

16. A drive shaft center support bearing is found to be rusty. The bearing should be:

    A. sprayed with WD-40®.

    B. lubricated with 90 weight oil.

    C. lubricated with wheel bearing grease.

    D. removed and replaced.

17. When replacing the gear lube in a differential, be sure to do all of the following EXCEPT:

    A. drain the old gear lube into a pan.

    B. fill the differential until the new fluid is 1 inch below the fill hole.

    C. dispose of the old gear lube using a licensed recycler.

    D. use new gear lube as recommended by the OEM.

18. A vehicle owner complains that it is difficult to get a 4WD vehicle into and out of 4WD. Which of the following is the most likely cause?

    A. The transfer case is overfilled with gear lube.

    B. The shift linkage is damaged.

    C. The universal joints need to be lubricated.

    D. The drive shaft is bent.

19. After having a tire replaced on a vehicle, a customer complains that the ABS light stays ON. Which of the following is the most likely cause?

    A. The ABS light circuit is open.

    B. The tire has been installed backwards.

    C. The wheel was bent during tire mounting.

    D. The tire is the wrong size.

20. Technician A says that Supplemental Restraint System (SRS) electrical connectors are often yellow in color. Technician B says that SRS circuits must be allowed to power down for 10 seconds. Who is correct?

    A. A only

    B. B only

    C. Both A and B

    D. Neither A nor B

21. A power steering pump is being replaced on a vehicle. Technician A says all the power steering hoses should also be replaced. Technician B says, depending on the system, the power steering filter should be cleaned or replaced. Who is correct?

    A. A only

    B. B only

    C. Both A and B

    D. Neither A nor B

22. Technician A says that if a strut rod bushing has deteriorated, it is easier to replace the entire strut rod. Technician B says that strut rods themselves seldom fail. Who is correct?

    A. A only

    B. B only

    C. Both A and B

    D. Neither A nor B

23. Which of the following parts, as shown in the illustration above, is being inspected for wear?

    A. A shock absorber bushing

    B. A strut bushing

    C. A spring shackle

    D. A spring bushing

24. When driving slowly over irregular surfaces, a clunking noise is heard coming from the front end of a vehicle. Technician A says that the vehicle's sway bar bushings may be faulty. Technician B says that the coil springs may be weak. Who is correct?

    A. A only

    B. B only

    C. Both A and B

    D. Neither A nor B

25. A vehicle is experiencing excessive rear-end sway on turns. Which of the following is the most likely cause?

    A. A broken sway bar link

    B. A faulty rear universal joint

    C. A faulty rear wheel bearing

    D. A broken axle flange

26. Technician A says that a faulty rear ball joint could affect a vehicle's rear wheel alignment. Technician B says that a faulty rear ball joint could be a safety hazard. Who is correct?

    A. A only

    B. B only

    C. Both A and B

    D. Neither A nor B

27. A mist of oil is seen on the tubes of a vehicle's rear shock absorbers. The shocks:

    A. are normal.

    B. will need to be replaced with in a few thousand miles.

    C. should be replaced as soon as possible.

    D. should be replaced immediately.

28. With the steering wheel straight, a vehicle tends to drift to the left while travelling on a level road. Technician A says that the air pressure in the vehicle's right front tire may be low. Technician B says that the vehicle may need a wheel alignment. Who is correct?

    A. A only

    B. B only

    C. Both A and B

    D. Neither A nor B

29. When a vehicle is being test driven straight on a level road, the steering wheel is not centered. Which of the following is the most likely cause?

    A. The clock spring is faulty.

    B. The steering wheel is incorrectly installed.

    C. The steering wheel is bent.

    D. A front strut is worn.

30. A vehicle experiences "dog tracking." Which of the following is the most likely cause?

    A. Rear axle misalignment

    B. Worn rear shocks

    C. Worn rear springs

    D. Worn rear tires

31. When driven at very slow speeds, a vehicle's steering wheel tends to shimmy. Which of the following is the most likely cause?

    A. A faulty shock

    B. A faulty spring

    C. A faulty control arm bushing

    D. A faulty tire

32. Technician A says that it's OK to repair a tire when two puncture repairs overlap each other. Technician B says when two tire punctures are within several inches of each other and on the same tread line, the tire should be scrapped. Who is correct?

    A. A only

    B. B only

    C. Both A and B

    D. Neither A nor B

33. During a test drive, a high-pitched metallic scraping noise is heard whenever a vehicle turns a corner. Which of these is the most likely cause?

    A. Faulty wheel bearings

    B. Faulty CV joints

    C. Worn brake pads

    D. Worn ball joints

34. A vehicle's red brake warning light stays on whenever the ignition is ON and the engine is running. Any of the following could be the cause EXCEPT:

    A. The parking brake is engaged.

    B. The brake fluid is low in the reservoir.

    C. The parking brake switch is out of adjustment.

    D. The ABS has a faulty wheel sensor.

35. Technician A says it's a good idea to open a caliper bleed screw before retracting the caliper piston. Technician B says to start brake bleeding at the wheel closest to the master cylinder. Who is correct?

    A. A only

    B. B only

    C. Both A and B

    D. Neither A nor B

36. Technician A says that whenever drum brake shoes are being replaced, the drum brake hardware should be cleaned in a parts washer and re-used. Technician B says that when installing duo-servo brakes, the larger of the two shoes should be installed towards the front of the vehicle. Who is correct?

    A. A only
    B. B only
    C. Both A and B
    D. Neither A nor B

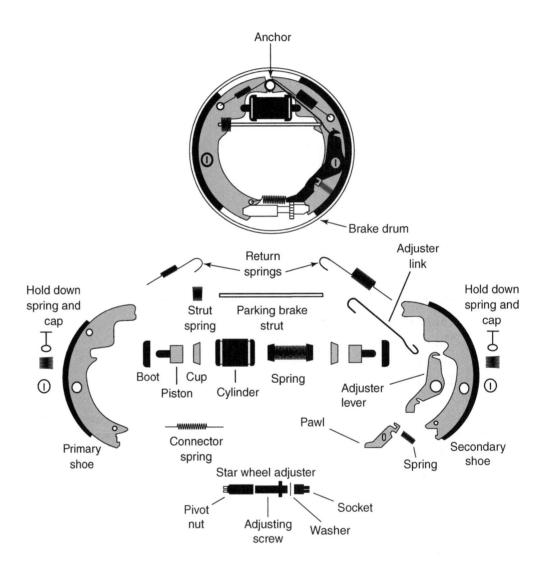

37. Which of the parts as shown in the illustration above, should be lubricated with water resistant grease during drum brake service?

    A. The parking brake strut and the adjuster lever
    B. The strut spring, connector spring and return springs
    C. The brake cylinder cups and boots
    D. The hold down spring and cap

38. A vehicle's parking brake should be adjusted:

    A. before the service brakes are adjusted.

    B. after the service brakes are adjusted.

    C. with the parking brake lever applied.

    D. with the ignition turned ON.

39. Technician A says that some disc brake pistons may be retracted using a "C" clamp. Technician B says that some caliper pistons may be retracted using a special retracting tool. Who is correct?

    A. A only

    B. B only

    C. Both A and B

    D. Neither A nor B

40. Upon removal of a vehicle's left-front brake pads, one pad in the floating-type caliper is worn much farther than the other. Which of the following is the most likely cause?

    A. A partially plugged brake hose

    B. Rusted caliper pins or slides

    C. A rusty caliper anchor plate

    D. A sticking caliper piston

41. Which of the following is used to check for rotor runout?

    A. An inside micrometer

    B. An outside micrometer

    C. A dial indicator

    D. A Vernier caliper

42. When refilling a brake master cylinder, always use:

    A. DOT 3 brake fluid.

    B. DOT 5 brake fluid.

    C. Synthetic brake fluid.

    D. OEM specified brake fluid.

43. A vehicle which uses the same hydraulic pressure pump for both the power brakes and power steering has a:

    A. Bendix® brake system.

    B. Hydro-boost brake system.

    C. Vacu-boost brake system.

    D. Teves® brake system.

44. Technician A says that current can be measured in a live circuit by connecting an ammeter in parallel across a load. Technician B says that current can be measured in a non-live (dead) circuit by using an amps clamp. Who is correct?

    A. A only

    B. B only

    C. Both A and B

    D. Neither A nor B

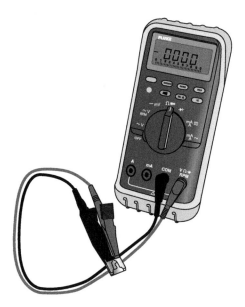

45. As shown in the illustration above, a fuse is being checked for continuity with a DMM. In this case, the fuse:

    A. is blown and should be replaced.

    B. has high resistance and should be replaced.

    C. has low resistance and should be cleaned.

    D. has no resistance and is good as is.

46. A battery load test is being performed. The proper method is to:

    A. apply a load equal to twice the CCA rating.

    B. apply a load equal to one-half the CCA rating.

    C. apply a load for 30 seconds.

    D. apply a load until the battery drops to 9.6 volts.

47. A 12-volt lead acid battery is being charged. Technician A says that when connecting the battery charger to a battery the cell caps should remain on the battery. Technician B says that when connecting the battery charger to the battery safety glasses should be worn. Who is correct?

    A. A only

    B. B only

    C. Both A and B

    D. Neither A nor B

48. A load test is being performed on a vehicle's charging system. Technician A says it's OK to increase the load until the charging system voltage just starts to drop, then take note the amperage. Technician B says it's OK to apply a load until the alternator light comes on, then take note of the amperage. Who is correct?

    A. A only

    B. B only

    C. Both A and B

    D. Neither A nor B

49. The puller shown above is used to remove a(an):

    A. ball joint.
    B. tie rod end.
    C. alternator decoupling pulley.
    D. steering knuckle.

50. A headlight alignment procedure is being performed. Which of the following is LEAST LIKELY to cause the alignment to be incorrect?

    A. A sloping drive surface
    B. Low tire pressure
    C. dirty headlight lenses
    D. A heavy load in the vehicle trunk

51. Technician A says that the brake warning light on the instrument panel cluster dash is a yellow light. Technician B says that the ABS indicator light on the IPC is red. Who is correct?

    A. A only
    B. B only
    C. Both A and B
    D. Neither A nor B

52. With the vehicle engine running, the A/C compressor fails to engage when the A/C switch is turned ON. Which of the following is the most likely cause?

   A.   A faulty blower motor

   B.   A faulty A/C relay

   C.   A faulty blend door sensor

   D.   A faulty heater core

53. During an underhood inspection for an inoperative A/C system, a technician notices an oily residue on the fins of the cooling system radiator. Which of the following could be the cause?

   A.   A faulty compressor

   B.   A faulty condenser

   C.   A faulty evaporator

   D.   A faulty expansion valve

54. Technician A says that leaves caught in the condenser could cause the cabin heating system to operate poorly. Technician B says that leaves caught in the condenser could cause the A/C system to operate poorly. Who is correct?

   A.   A only

   B.   B only

   C.   Both A and B

   D.   Neither A nor B

55. Technician A says that the best way to determine if a newer style serpentine belt is worn is to visually inspect it. Technician B says that the best way to determine if a newer style serpentine belt is worn is to check it with a wear gauge. Who is correct?

   A.   A only

   B.   B only

   C.   Both A and B

   D.   Neither A nor B

PREPARATION EXAM 2

1. A high mileage engine has an leaking oil pan gasket. Technician A says some undercar components may need to be removed in order to replace the pan gasket. Technician B says that the engine may have to be raised up off its engine mounts to replace the gasket. Who is correct?

   A.   A only

   B.   B only

   C.   Both A and B

   D.   Neither A nor B

2. Oil pan and valve cover gasket(s) should be checked for:

   A. rust.

   B. stains.

   C. corrosion.

   D. leaks.

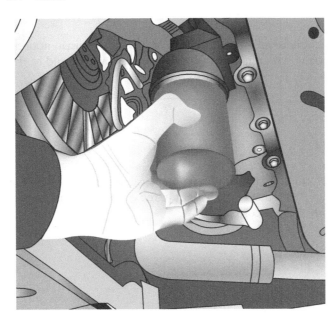

3. The technician shown in the illustration above is changing the:

   A. gasoline filter.

   B. transmission filter.

   C. differential filter.

   D. oil filter.

4. Which of the following is the preferred method of checking a newer style serpentine belt for wear?

   A. Twist the belt one-quarter turn and check for tightness.

   B. Use a belt tension gauge.

   C. Use a serpentine belt wear gauge.

   D. Check the wear indicator on the pulley.

5. A vehicle's cooling system has been drained and flushed, and fresh coolant is being added to a vehicle. Technician A says to mix the correct ratio of coolant to water for the correct freeze point. Technician B says a bleed screw may have to be opened to purge air from the cooling system. Who is correct?

   A. A only

   B. B only

   C. Both A and B

   D. Neither A nor B

6. Any routine inspection and maintenance program includes the periodic replacement of the:

   A. transmission filter.

   B. power steering fluid filter.

   C. cabin filter.

   D. air filter.

7. Which of the following should be used when replacing defective vehicle exhaust system components?

   A. Galvanized exhaust system components

   B. Aluminum exhaust system components

   C. Stainless steel exhaust system components

   D. Titanium exhaust system components

8. When checking for diagnostic trouble codes (DTCs), it's important to:

   A. use the proper scan tool and software for the vehicle under test.

   B. clear all the codes before performing any tests.

   C. check the vehicle owner's manual before performing any tests.

   D. take the vehicle for a test drive to run all the monitors.

9. On an OBD II vehicle, an EVAP leak may be associated with any of the following events EXCEPT:

   A. the MIL will be illuminated.

   B. a rubber hose may be split.

   C. carbon monoxide will be released to the atmosphere.

   D. the gas cap will allow VOCs to leak to the atmosphere.

10. Which of the following is the LEAST LIKELY way a technician would confirm that an automatic transmission is faulty?

   A. Run the engine with the vehicle in NEUTRAL.

   B. Interview the customer.

   C. Take the vehicle for a test drive.

   D. Check for fault codes in the TCM.

11. Technician A says that the technician shown in the illustration above, could be checking the differential fluid level. Technician B says that the technician shown in this illustration could be checking the transmission fluid level. Who is correct?

A. A only

B. B only

C. Both A and B

D. Neither A nor B

12. A technician is inspecting a vehicle's rear transmission mount and finds its rubber is dry and has split. The mount should be:

A. replaced it with a solid steel mount.

B. replaced with a synthetic mount.

C. replaced with similar used part.

D. replaced with a new OEM or aftermarket part.

13. Technician A says that all automatic transmission fluids (ATF) are interchangeable. Technician B says that new ATF should be pink in color. Who is correct?

A. A only

B. B only

C. Both A and B

D. Neither A nor B

14. Following drivetrain repairs to a vehicle, a final test drive reveals a vibration problem which increases with road speed and engine torque. Which of the following is the most likely cause?

   A.   Incorrect drive line phasing

   B.   Over-torqued wheel lug nuts

   C.   Excessive pinion gear lash

   D.   Loose spring shackles

15. While a brake shoe replacement procedure is being performed, gear lube is seen leaking from behind the rear axle flange. Technician A says that a leaking axle seal could be the cause. Technician B says that a faulty brake drum seal could be the cause. Who is correct?

   A.   A only

   B.   B only

   C.   Both A and B

   D.   Neither A nor B

16. A wheel stud needs to be replaced on a solid rear axle flange. Technician A says that on some vehicles the axle may have to be removed to make installing the replacement wheel stud possible. Technician B says that on some vehicles it may be possible to install a new wheel stud without removing the axle. Who is correct?

   A.   A only

   B.   B only

   C.   Both A and B

   D.   Neither A nor B

17. A FWD vehicle experiences a clicking noise from the left front wheel area during slow left-hand turns. Which of the following is the most likely cause?

   A.   A noisy brake pad

   B.   Worn ball joints

   C.   Worn sway bar bushings

   D.   A faulty CV joint

18. A sealed front wheel bearing is to be installed. Technician A says the hub may have to be removed in order to install the bearing. Technician B says the bearing may have to be packed before installation. Who is correct?

   A.   A only

   B.   B only

   C.   Both A and B

   D.   Neither A nor B

19. Which of the following is true about the fluid used in a transfer case?

   A.   It is the same as the gear lube used in the differential.

   B.   It may be unique because of specially designed additives.

   C.   It is the same as automatic transmission fluid (ATF).

   D.   It is the same as that used in manual transmissions.

20. Any of the following would likely indicate the correct power steering fluid to be used in a vehicle EXCEPT:

    A. the owner's manual.

    B. the service manual.

    C. a parts and labor guide.

    D. a service bulletin.

21. A vehicle exhibits a groaning noise whenever the steering wheel is turned. Which of the following is the most likely cause?

    A. The system is overfilled with fluid.

    B. The system needs to be purged of air.

    C. The power steering pump is faulty.

    D. The power steering rack is faulty.

22. During an undercar inspection, a bent lower control arm is discovered. Technician A says it may be heated and straightened. Technician B says that after the appropriate repair, the wheels should be realigned. Who is correct?

    A. A only

    B. B only

    C. Both A and B

    D. Neither A nor B

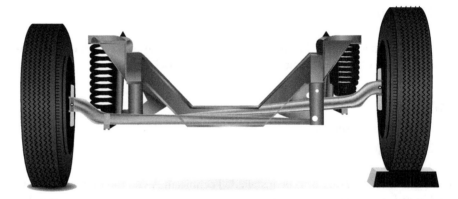

23. An older model pickup with a twin I-beam front suspension as shown above is found to need a camber adjustment. Technician A says that the I-beam(s) may need to be bent to achieve the alignment. Technician B says the radius rods may need to be bent to achieve the alignment. Who is correct?

    A. A only

    B. B only

    C. Both A and B

    D. Neither A nor B

24. The front-end coil springs on a vehicle have sagged. Which of the following would be most affected?

    A. Caster
    B. Toe
    C. Thrust line
    D. Ride height

25. The rear coil spring on a vehicle is found to be broken. Which of the following is the LEAST LIKELY cause?

    A. The vehicle was overloaded.
    B. The vehicle was used for off-roading.
    C. The vehicle was not well maintained.
    D. The vehicle was used for heavy towing.

26. The rear rebound bumpers of a vehicle are found to be cracked and collapsed. Technician A says that routine vehicle overloading could be the cause. Technician B says that once the bumpers are replaced, a wheel alignment will be needed. Who is correct?

    A. A only
    B. B only
    C. Both A and B
    D. Neither A nor B

27. Upon inspection, the rear tie rods of an independently suspended vehicle are found to have 1/4 inch of play. Technician A says that they should be replaced. Technician B says that if tie rods are replaced, a wheel alignment will be required. Who is correct?

    A. A only
    B. B only
    C. Both A and B
    D. Neither A nor B

28. A vehicle tends to pull to one side of the road when travelling on a straight road. Which of the following is the most likely cause?

    A. Incorrect steering wheel indexing/installation
    B. Incorrect thrust line
    C. Incorrect camber setting
    D. A faulty steering damper

29. Technician A says that caster on a short-long arm (SLA) equipped vehicle may be adjusted by adding or subtracting shims. Technician B says that caster on a SLA equipped vehicle may be adjusted by rotating an eccentric shaft. Who is correct?

    A. A only
    B. B only
    C. Both A and B
    D. Neither A nor B

30. A tire is found to have a nail hole in its sidewall. Which of the following should be done?

    A. Repair the tire with a string type plug.

    B. Repair the tire with a patch from the inside.

    C. Repair the tire with a vulcanizing kit.

    D. Scrap and replace the tire with a similar good one.

31. The four non-asymmetric/non-directional radial tires on a 4WD vehicle with a mini-spare tire are to be rotated. Which of the following is the correct procedure?

    A. Exchange both front and rear tires from side to side.

    B. Exchange the front and rear tires with each other on the same side.

    C. Move the rear tires to the front and the front tires to the opposite rear.

    D. Move the rear tires to the opposite front and the front tires to the rear.

32. A vehicle with the same make/size of tires and wheels and the same recommended inflation on all 4 wheels experiences an illuminated TPMS light. Technician A says that the vehicle's tires may have been rotated. Technician B says that a TPMS relearn procedure was not performed. Who is correct?

    A. A only

    B. B only

    C. Both A and B

    D. Neither A nor B

33. The yellow ABS warning light is illuminated on a vehicle's instrument panel. Which of the following is the most likely cause?

    A. A wheel is out of balance.

    B. A tire is defective.

    C. A TPMS sending unit is faulty.

    D. A tone wheel was damaged during a tire repair.

34. A short to ground in the parking light circuit will likely cause:

    A. the brake warning light to stay on.

    B. the brake warning light to not illuminate.

    C. an MIL to be illuminated.

    D. a code to be set in the EBCM.

35. The lid on a can of DOT-3 brake fluid has been left off its container overnight. Technician A says it's OK to replace the cap and continue to use the fluid. Technician B says the brake fluid may be filtered through cheese cloth and then used. Who is correct?

    A. A only

    B. B only

    C. Both A and B

    D. Neither A nor B

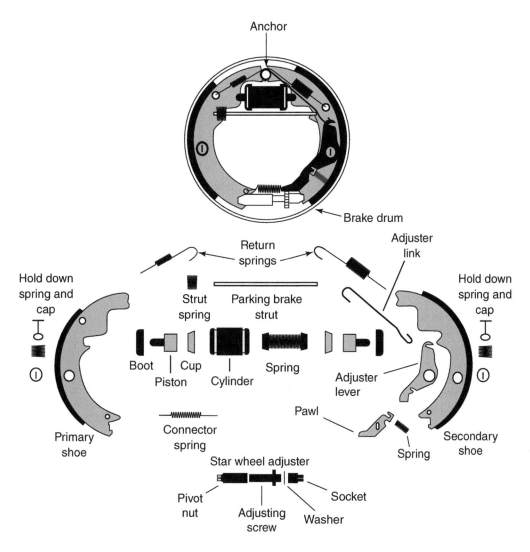

36. Brake shoes are being installed at the rear of a vehicle. Which of the parts in the illustration above, should be lubricated with water resistant grease?

   A.   The brake cylinder cups

   B.   The hold downs

   C.   The shoe linings

   D.   The star wheel adjuster

37. A technician is attempting to retract a brake caliper piston by using a "C" clamp, but the piston will not retract. Which of the following is the most likely cause?

   A.   The caliper piston also serves as a parking brake.

   B.   The caliper piston "O" ring seal is twisted.

   C.   The caliper is a full floating type component.

   D.   the caliper is a semi-floating type.

38. While inspecting the disc brakes of a vehicle, the technician finds that only the inner pad has worn down to its backing plate. Technician A says that the caliper piston should be inspected for possible damage from the pad. Technician B says that the rotor should be inspected for damage from the pad. Who is correct?

    A. A only

    B. B only

    C. Both A and B

    D. Neither A nor B

39. During a routine disc brake inspection, the caliper slide pin boots are found to be torn. Which of the following should be done?

    A. Replace the caliper pins and boots.

    B. Inspect the caliper pins for corrosion.

    C. Inspect the caliper pistons.

    D. Replace the caliper slides.

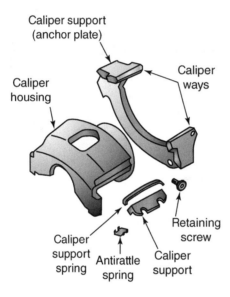

40. The disc brake components illustrated above are being assembled. Which of the following part(s) should be lubricated with waterproof grease?

    A. The caliper housing

    B. The caliper ways

    C. The support spring

    D. The caliper anchor plate

41. Which of the following is the best method to use when checking a vacuum brake booster check valve for leaks?

    A. Remove the valve and blow compressed air through it to see if it leaks.

    B. Run the engine, then shut it off and pull the brake booster check valve to see if air rushes in.

    C. Remove the valve and see if light passes through it.

    D. Remove the vacuum line and attempt to blow air through it.

42. An electric-hydraulic assist brake system is to be checked for leaks. Technician A says that the hydraulic hoses should be inspected for signs of leakage. Technician B says that the hose fittings should be inspected for signs of leakage. Who is correct?

    A. A only
    B. B only
    C. Both A and B
    D. Neither A nor B

43. A hydraulic type brake booster system is to be inspected for faults. The inspection should start with all of the following EXCEPT:

    A. the brake fluid level.
    B. the power steering belt.
    C. the hose connections.
    D. the master cylinder pushrod.

44. An air bag system (supplemental restraint system - SRS) needs to be disarmed before SRS related service is performed. Technician A says to pull the SRS fuse. Technician B says to disconnect the battery. Who is correct?

    A. A only
    B. B only
    C. Both A and B
    D. Neither A nor B

45. When measuring for load resistance in a circuit, the technician should be sure to:

    A. turn the circuit power off.
    B. have the circuit power on.
    C. connect the ohmmeter in series with the load.
    D. connect the ammeter in parallel with the load.

46. Before disconnecting a vehicle's 12-volt battery for service, providing auxiliary 12-volt power to the vehicle will accomplish all of these EXCEPT:

    A. make certain engine operating parameters are maintained in the PCM.
    B. pose the risk of damage to the PCM.
    C. maintain automatic power seat-position memory settings.
    D. keep radio station presets stored in keep alive memory (KAM).

47. Which of the following is the best procedure to follow when using a battery charger on a discharged battery?

    A. Use a fast charge with high amperage on flooded lead-acid batteries.
    B. Use a slow charge with high amperage on flooded lead-acid batteries.
    C. Use a slow charge with high voltage on valve regulated lead-acid batteries.
    D. Use the recommended volts/amps charge rate specified for the battery type.

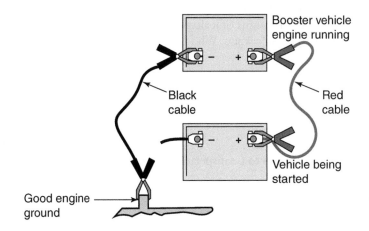

48. When attempting to jump-start a vehicle illustrated above, which of the following jumper cable connections should be made last?

A. Red cable to the booster vehicle

B. Red cable to the vehicle being started

C. Black cable to the booster vehicle

D. Black cable to a good engine ground

49. A starter motor is being removed from a vehicle. Technician A says it's important to disconnect the heavy battery wire from the starter solenoid and tape it to prevent a short circuit to a vehicle ground. Technician B says it's important to remove and tape the small wire lead to the starter solenoid to prevent it from shorting to a vehicle ground. Who is correct?

A. A only

B. B only

C. Both A and B

D. Neither A nor B

50. An alternator pulley designed to smooth out crankshaft torque reversals otherwise imposed on the alternator is to be replaced. Which of the following should be used?

A. Solid pulley

B. Overrunning pulley

C. Decoupling pulley

D. Overdrive pulley

51. A heater control panel with an LED display fails to illuminate and the controls do not operate properly. Technician A says to check the fuse and all electrical connections. Technician B says the circuit board or even the entire panel may have to be replaced. Who is correct?

A. A only

B. B only

C. Both A and B

D. Neither A nor B

52. Moving the heater control switch knob from HEAT to DEFROST causes a vehicle's A/C compressor to engage and the underhood electric fan to start. Which of the following could be the cause?

    A. A grounded compressor clutch circuit
    B. A faulty heater control switch
    C. A faulty fan circuit
    D. The A/C circuit is working correctly

53. Which of the following would happen if the A/C condenser becomes blocked with leaves and debris?

    A. The A/C system would blow too much cold air.
    B. The high-side refrigerant pressure in the A/C system would rise.
    C. The evaporator would frost up.
    D. The condenser would frost up.

54. A vehicle's serpentine belt squeals whenever the A/C is turned to ON. Technician A says the compressor may be locked up. Technician B says the serpentine belt may be loose. Who is correct?

    A. A only
    B. B only
    C. Both A and B
    D. Neither A nor B

55. With the A/C ON, water is seen blowing from the A/C vents when the blower motor is turned to HIGH. Which of the following is the most likely cause?

    A. A leaking A/C condenser
    B. A leaking expansion valve
    C. A plugged heater core
    D. A plugged condensate drain

PREPARATION EXAM 3

1. An oil pan is to be removed and replaced. Which of the following is LEAST LIKELY to be used to seal the oil pan when it is reinstalled?

    A. a cork gasket
    B. a synthetic gasket
    C. silicone-based sealant
    D. RTV sealant

2. All of the following cooling system components should be periodically inspected and pressure tested EXCEPT for the:

    A. radiator.
    B. evaporator.
    C. heater hoses.
    D. pressure cap.

3. A radiator hose is found to be soft and "squishy" when squeezed. Technician A says the hose has aged and may fail before too long. Technician B says that radiator hoses are sometimes uniquely shaped and replacement hoses will only fit on certain vehicle models. Who is correct?

    A. A only
    B. B only
    C. Both A and B
    D. Neither A nor B

4. When a cooling system thermostat fails open it usually will:

    A. not allow the engine to warm up.
    B. cause the engine to overheat.
    C. cause the cabin heater core to malfunction.
    D. not allow the A/C to operate.

5. On vehicles with a V-belt-driven fan, which of the following components drives the thermostatic clutch?

    A. The camshaft
    B. The crankshaft
    C. The balance shaft
    D. The intermediate shaft

6. An illuminated engine oil warning light on the dashboard usually indicates that the:

    A. oil is too hot.
    B. oil pressure is too low.
    C. oil pressure relief valve is stuck closed.
    D. oil pressure switch is stuck open.

7. All of the following are routinely checked during a 12,000 mile maintenance inspection EXCEPT for the:

    A. air filter.
    B. oil level.
    C. tire pressures.
    D. cabin filter.

8. Which of the following is best for retrieving engine-related diagnostic trouble codes (DTCs) on an OBD II vehicle?

    A. A digital multimeter
    B. An analog voltmeter
    C. A lab-type oscilloscope
    D. A scan tool

9. Shortly after the gas tank is filled, an MIL illuminates. Which of the following should the technician do first?

    A. Pull any trouble codes in the PCM.
    B. Take a look at freeze frame data in the PCM.
    C. Check under the hood for loose or faulty EVAP hoses.
    D. Check if the gas cap is loose.

10. Type "F" automatic transmission fluid (ATF) is used:

    A. for most automatic transmission vehicles.
    B. only for certain Ford® transmissions.
    C. only for Ferrari® transmissions.
    D. for vehicles made in Korea.

11. An automatic transmission experiences a mild leak of trans fluid at the oil pan gasket. Technician A says that leak-stopping additive may be added to stop the leak. Technician B says to use RTV sealant at the leak area to stop the leak. Who is correct?

    A. A only
    B. B only
    C. Both A and B
    D. Neither A nor B

12. A driveshaft universal joint (U-joint) is dry of lubricant and found to be loose. Which of the following is the proper repair?

    A. Grease the U-joint.
    B. Tighten the U-joint.
    C. Grease and tighten the U-joint.
    D. Replace the U-joint.

13. A transmission mount is being inspected for wear. Technician A says to look for oil-soaked and swelled rubber between the mounting plates. Technician B says to look for separated, torn, or collapsed rubber between the mounting plates. Who is correct?

    A. A only
    B. B only
    C. Both A and B
    D. Neither A nor B

14. A manual transmission vehicle is hard to shift. Technician A says that the shift linkage may be misaligned. Technician B says the hydraulic clutch reservoir may be overfilled. Who is correct?

   A. A only
   B. B only
   C. Both A and B
   D. Neither A nor B

15. A manual transmission's fluid level is being drained and refilled. Technician A says that some manual transmissions use the same type fluid as that used in engine crankcases. Technician B says that some manual transmissions use the same type fluid as that used in differentials. Who is correct?

   A. A only
   B. B only
   C. Both A and B
   D. Neither A nor B

16. A driveshaft on a front wheel drive (FWD) vehicle with McPherson struts must be removed in order to replace an alternator. Technician A says the tie rod will first need to be disconnected from the steering knuckle. Technician B says that if a tie rod end is disconnected, a wheel alignment will be needed to correct the toe setting. Who is correct?

   A. A only
   B. B only
   C. Both A and B
   D. Neither A nor B

17. A few months after a front wheel bearing has been replaced in a rear wheel drive (RWD) pickup, the bearing is found to be making noise and the outer race it is found to be badly scored and wearing unevenly. Which of the following is the most likely cause of the premature bearing failure?

   A. An incorrect part number bearing was installed.
   B. The axle nut was improperly torqued.
   C. A small metal burr kept the bearing from properly seating.
   D. The lugs nuts were improperly torqued.

18. Transfer case fluid in a 4WD vehicle is to be drained and replaced. Technician A says that some transfer cases have the word "Drain" cast into the housing to indicate the drain plug location. Technician B says that some drain plugs can be removed by using a 3/8" extension on a ratchet wrench. Who is correct?

   A. A only
   B. B only
   C. Both A and B
   D. Neither A nor B

19. All of the following statements about replacing a leaking transaxle half-shaft seal in a 4WD transverse engine vehicle are true EXCEPT the:

    A. axle seal can be replaced without removing the half shaft.

    B. half-shaft will have to be removed in order to replace the seal.

    C. steering knuckle must be disconnected from the steering linkage.

    D. large half-axle nut is easiest to loosen if done before the vehicle is raised from the shop floor.

20. Diagnostic trouble codes (DTCs) are to be retrieved on a vehicle. Technician A says that on some vehicles, simply turning the ignition ON-OFF-ON three times in succession will cause DTCs to display on the electronic odometer. Technician B says that some vehicles' DTCs can be read on the electronic climate control panel. Who is correct?

    A. A only

    B. B only

    C. Both A and B

    D. Neither A nor B

21. Power steering fluid is seen dripping from the power steering return hose where it connects to the PS reservoir. Which of the following is the most likely cause?

    A. The fluid is the wrong viscosity.

    B. The fluid is the wrong type.

    C. The banjo fitting washer is split.

    D. The hose clamp needs to be tightened.

22. Which of the following is illustrated above?

    A. A CV joint boot

    B. A rack and pinion boot

    C. A strut boot

    D. A shift lever boot

23. Worn ball joints can best be checked for looseness and wear when the vehicle is:

    A. fully loaded and bounced up and down.

    B. empty and rocked from side to side.

    C. raised slightly off the ground and each wheel is forced to move up and down.

    D. raised slightly off the ground and each wheel is rotated.

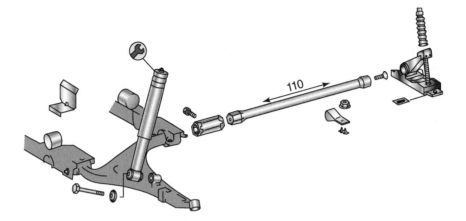

24. What is the purpose of suspension part number 110 in the illustration above?

    A. It prevents vehicle body sway.

    B. It serves to dampen body roll.

    C. It serves to soften the vehicle ride.

    D. It serves to dampen body yaw.

25. The rear ride height on a live rear-axle pickup truck becomes too low (sags) when normal tools and equipment are being transported. Which of the following is the most likely cause?

    A. Weak struts

    B. Weak shocks

    C. Weak lateral/track bar

    D. Weak leaf springs

26. The rear struts are being inspected on a sports sedan and one is found to be leaking. Technician A says both rear struts should be replaced. Technician B says a front-end wheel alignment will be necessary after they are replaced. Who is correct?

    A. A only

    B. B only

    C. Both A and B

    D. Neither A nor B

27. A vehicle experiences excessive rear-end sway. Technician A says that the rear knuckle bushings may be worn. Technician B says that if a knuckle bushing goes bad, the entire knuckle will need to be replaced. Who is correct?

    A. A only
    B. B only
    C. Both A and B
    D. Neither A nor B

28. Technician A says that the object shown in the illustration above is used for spreading a tire open when the tire is being patched. Technician B says that the object shown is for holding the steering wheel stationary and centered during a wheel alignment. Who is correct?

    A. A only
    B. B only
    C. Both A and B
    D. Neither A nor B

29. If a camber adjustment on a McPherson strut front end is to be performed, the:

    A. upper strut bearing may be moved fore or aft.
    B. upper strut bearing may be moved inward or outward.
    C. lower ball joint may be moved for or aft.
    D. lower ball joint may be moved inward or outward.

30. A rear wheel toe adjustment:

    A. is not normally possible on a live rear-axle vehicle.

    B. may be performed at the tie rods of a live rear-axle vehicle.

    C. is not possible on a vehicle with an independently suspended rear end.

    D. may be performed on a vehicle with a solid rear axle.

31. When driving straight, the steering wheel of a vehicle is found to be off center. Technician A says that resetting the steering wheel sensor will address the concern. Technician B says a wheel alignment will address the concern, but following the alignment, the steering wheel sensor will need to be reset. Who is correct?

    A. A only

    B. B only

    C. Both A and B

    D. Neither A nor B

32. What is the purpose of the device shown above?

    A. It is used when patching a tire.

    B. It is used for checking tire pressures.

    C. It is used for installing TPMS sensors.

    D. It is used for checking tire tread depth.

33. Which of the following is the LEAST LIKELY method to be used when dynamically balancing an out-of-balance tire and wheel assembly?

    A. Fasten a weighted plate to the wheel rim.

    B. Clamp a lead weight to the lip of a steel wheel rim.

    C. Tape a lead weight to the inner rim of a custom wheel rim.

    D. Clamp a zinc weight to the lip of the steel wheel rim.

34. While attempting to replace a leaking wheel cylinder, a technician kinks the brake line near the wheel cylinder. Technician A says that new brake line steel tubing may be purchased, cut to length, double-flared and correctly shaped as a replacement. Technician B says it's possible to purchase a straight brake line with flared ends and the correct fittings installed at each end as a replacement. Who is correct?

    A. A only
    B. B only
    C. Both A and B
    D. Neither A nor B

35. Technician A says a hand-held vacuum pump may be used to help bleed a brake system. Technician B says a pressure bleeder requires two people to use. Who is correct?

    A. A only
    B. B only
    C. Both A and B
    D. Neither A nor B

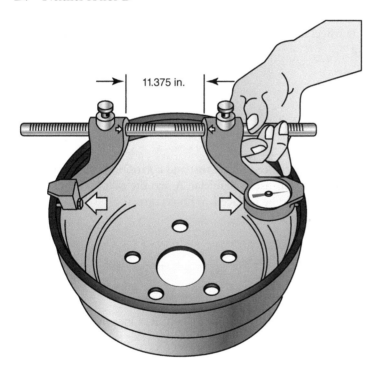

11.375 in.

36. Technician A says the tool in the illustration above is being used to measure a brake drum's diameter so that the brake shoes may be pre-adjusted. Technician B says the drum is being checked for wear beyond its usable limit. Who is correct?

    A. A only
    B. B only
    C. Both A and B
    D. Neither A nor B

37. Which of the following tools should be used when removing a brake line fitting from a wheel cylinder?

    A. An open-end wrench

    B. A box-end wrench

    C. A flare wrench

    D. A combination wrench

38. A vehicle is to be raised on a lift to have its parking brake system inspected for proper operation. Technician A says that, on a drum-type parking brake system, to inspect the parking brake cables, equalizer, and self-adjusting linkage. Technician B says a caliper-type parking brake system should have its cables, equalizer, cams, and screw mechanisms inspected. Who is correct?

    A. A only

    B. B only

    C. Both A and B

    D. Neither A nor B

39. It is important to torque wheel lug nuts or bolts to OEM specs, and in the correct sequence, in order to help prevent:

    A. brake piston binding.

    B. disc warpage.

    C. brake squeal.

    D. cracking the drums.

40. A vehicle's brake calipers have experienced severe rust and corrosion both from exposure to a salt water environment and from non-use. Technician A says the calipers should be rebuilt. Technician B says the calipers should be replaced. Who is correct?

    A. A only

    B. B only

    C. Both A and B

    D. Neither A nor B

41. Warped rotors are being replaced with new ones on a vehicle. Once installed, the new rotors will need to be:

    A. measured for runout.

    B. checked for warpage.

    C. sanded to remove residual rust.

    D. burnished, or "burned in."

42. What happens if off-car brake drum machining is performed too fast on a brake lathe, with only a rough cut and no fine-finish cut made?

   A. Once in use, the brakes are likely to make a squealing noise.

   B. Once in use, the brake shoes will overheat.

   C. The shoes will "click" as they attempt to follow spiral grooves in the drum.

   D. The brakes are likely to make a humming noise.

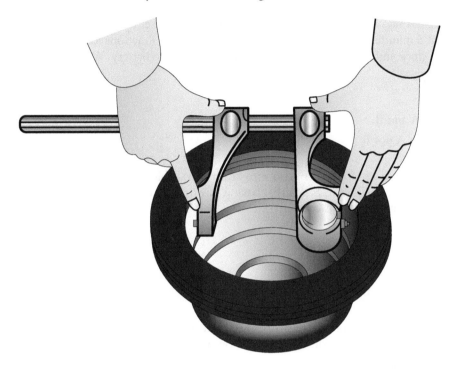

43. What action is taking place in the illustration above?

   A. A brake drum is being measured for wear.

   B. A parking brake drum's inside diameter is being measured.

   C. A brake rotor's circumference is being measured.

   D. A brake rotor is being checked for warpage.

44. A power brake booster is being checked for proper operation. Technician A says if when the engine is run and then turned off, the brake pedal is pumped until the pedal feels harder and higher, and the engine is re-started, the brake pedal should drop slightly upon engine restart. Technician B says to first start the engine, push and hold the brake pedal down, turn off the engine and pump the pedal to see if the brake pedal feels harder and rises. Who is correct?

   A. A only

   B. B only

   C. Both A and B

   D. Neither A nor B

45. While measuring for current in a circuit, a DMM reads "0.035". Which of the following is another way a stating the DMM's reading?

   A.   35 micro amps

   B.   35 milliamps

   C.   3.5 amps

   D.   35 amps

46. Technician A says that keep alive memory (KAM) can be reset by disconnecting the battery for about 5 minutes. Technician B says that KAM memory functions for the radio, seat position, clock and other settings prior to disconnecting the battery. Who is correct?

   A.   A only

   B.   B only

   C.   Both A and B

   D.   Neither A nor B

47. Technician A says a specially designed wire brush is available to clean the terminals of a side-terminal battery. Technician B says that baking powder can be used to neutralize corrosion on battery terminals. Who is correct?

   A.   A only

   B.   B only

   C.   Both A and B

   D.   Neither A nor B

48. A vehicle with a fully charged battery does not make a "click" sound or crank the engine when the ignition key is turned to the START position. Which of the following could be the cause?

   A.   A faulty neutral-safety switch

   B.   A faulty starter motor

   C.   A faulty automatic transmission clutch switch

   D.   A faulty battery

49. A charging system output test is to be performed by placing a high-amperage "amps clamp" on the heavy lead from the battery to the alternator. The clamp's "+" sign is pointing toward the battery. If the charging system is loaded and working properly, the digital reading would show:

   A.   high-amperage positive output from the alternator.

   B.   low-amperage positive output from the alternator.

   C.   high-amperage negative (minus) output from the alternator.

   D.   a high-amperage reading to the starter.

50. A rebuilt alternator is being installed in a vehicle. Which of the following could likely occur if a crowbar is used to tension the alternator when tightening the mounting bracket and serpentine drive belt?

   A.   The alternator fields could be damaged.

   B.   The alternator rotor may be damaged.

   C.   The mounting ear(s) could be broken off.

   D.   The drive belt could be snapped.

51. A vehicle's plastic headlight lenses are clouded enough to cause the vehicle to fail a safety inspection. Technician A says that the lenses can restored by using a fine grade of rubbing compound or toothpaste. Technician B says the lenses can be restored using a restoration kit containing fine grit sandpaper and a special liquid. Who is correct?

A. A only

B. B only

C. Both A and B

D. Neither A nor B

52. A vehicle's horn continues to blow whenever the ignition is switched to the ON position. Which of the following could be the cause?

A. A faulty ignition switch

B. A faulty horn

C. A faulty horn relay

D. An open in the horn circuit

53. Signs of oil seepage are seen at the A/C connecting block on a vehicle's firewall. Which of the following is the most likely cause?

A. A faulty connecting block

B. A faulty "O" ring at the connection

C. A broken fitting

D. A broken connection at the compressor

54. A vehicle's A/C condenser seems too hot, and overall A/C performance is poor. Which of the following is the most likely cause?

A. The A/C clutch is weak and slipping.

B. The evaporator is leaking refrigerant.

C. The condenser's airflow is blocked by debris.

D. The compressor is worn out.

55. A vehicle's cabin air filter is due for inspection. If found to be dirty, it would normally be:

A. blown free of debris with low pressure shop air and reinstalled.

B. washed with mild dish soap, rinsed, air dried and reinstalled.

C. cleaned of debris with a shop vacuum cleaner and reinstalled.

D. replaced.

PREPARATION EXAM 4

1. A vehicle engine is leaking oil from its bell housing. Technician A says the PCV valve should be inspected for being clogged. Technician B says the rear oil seal probably needs to be replaced. Who is correct?

A. A only

B. B only

C. Both A and B

D. Neither A nor B

2. An engine oil pan has been removed in order to replace a leaking gasket. At the same time the oil pan should be inspected for any of the following EXCEPT:

   A. dents.

   B. scratches.

   C. cracks.

   D. chips.

3. Modern car engines generally require oil changes at:

   A. 2,500 mile intervals.

   B. 3,000 mile intervals.

   C. 5,000 mile intervals.

   D. 10,000 mile intervals.

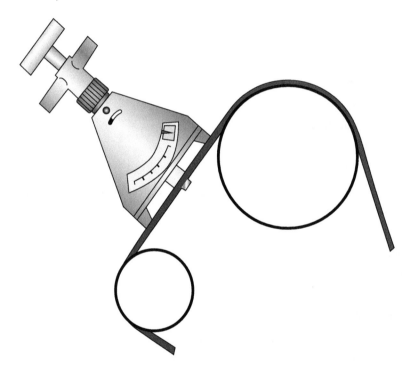

4. The tool in the illustration above is being used to check a:

   A. drive belt for wear.

   B. drive belt for correct tension.

   C. pulley for alignment.

   D. tensioner for alignment.

5. A cooling system is to be drained. Technician A says that in order to get all impurities out of the cooling system, it should be drained while hot. Technician B says in order to remove a radiator cap, it should be pushed down and turned counterclockwise. Who is correct?

   A. A only

   B. B only

   C. Both A and B

   D. Neither A nor B

6. Which of the following normally drives the thermostatic clutch on vehicles with a belt-driven fan?

    A. The camshaft

    B. The crankshaft

    C. The balance shaft

    D. The intermediate shaft

7. A routine 5,000 mile interval inspection includes checking the:

    A. transmission filter.

    B. power steering fluid filter.

    C. cabin filter.

    D. air filter.

8. The trouble codes on an OBD I Ford® are to be retrieved. Technician A says to use the self-test connector located under the hood on the driver's side of the vehicle. Technician B says to insert a jumper wire in the self-test connector and use a digital multimeter to count the needle swings and read the codes. Who is correct?

    A. A only

    B. B only

    C. Both A and B

    D. Neither A nor B

9. A 1996 model OBD II vehicle registered and driven in the "snowbelt" region of the country fails an emissions test. Which of the following is the most likely cause?

    A. A faulty fuel pump

    B. A cracked EVAP canister

    C. A rusted fuel filler neck

    D. A leaking fuel rail

10. The volatility of gasoline in a vehicle's tank is to be checked. Technician A says that testing the fuel's Reid Vapor Pressure will reveal the fuel's volatility. Technician B says that higher fuel volatility fuel is sold during the summer months. Who is correct?

    A. A only

    B. B only

    C. Both A and B

    D. Neither A nor B

11. Technician A says that a leaking transmission line may not be apparent because the fluid (ATF) could burn off before it drips from the leak to the shop floor. Technician B says that a transmission line leak would likely appear as red colored ATF. Who is correct?

    A. A only

    B. B only

    C. Both A and B

    D. Neither A nor B

12. Technician A says that powertrain mounts can be checked for defects by revving the engine in gear with the brakes firmly applied and watching how much the engine lifts up as it applies torque to the drive line. Technician B says that an engine mount can be inspected by jacking up the engine slightly and looking for an air gap where the mount's cushioning rubber should be. Who is correct?

    A. A only

    B. B only

    C. Both A and B

    D. Neither A nor B

13. After a transmission pan has been removed to replace the transmission filter, a large magnet is found in the pan with gooey grey matter adhering to it. Which of the following statements best describes why the magnet is in the transmission pan?

    A. The magnet was inadvertently left in the transmission pan during transmission assembly.

    B. The magnet is in the pan to collect dirt which finds its way into the transmission fluid.

    C. The magnet is in the pan to collect fine metal wear particles in the transmission fluid.

    D. The magnet is in the pan to collect large metal debris in the event of a transmission failure.

14. The column shift mechanism on an older pickup truck with a manual transmission is sloppy and won't allow the transmission to go into all gears. Which of the following is the most likely cause?

    A. The shaft attached to the column shift lever is bent.

    B. The column shift lever shaft retaining bolts behind the dashboard have become loose.

    C. The column shift lever shaft is broken.

    D. The shift cable is kinked.

15. A manual transmission driveshaft seal is to be replaced. Technician A says to unbolt the driveshaft and remove the U-joint yoke to get at the seal. Technician B says that if there is too much play or slop in the U-joint yoke, a new seal will likely leak shortly after it's installed. Who is correct?

    A. A only

    B. B only

    C. Both A and B

    D. Neither A nor B

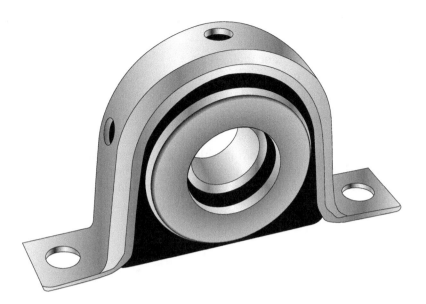

16. Which of the following driveshaft parts is pictured in the illustration above?

    A. Surface bearing
    B. Center support bearing
    C. Torrington bearing
    D. Thrust bearing

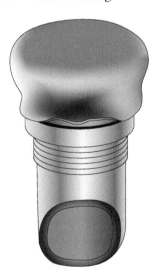

17. Which of the following parts is illustrated above?

    A. A tire valve
    B. An axle housing vent
    C. A TPMS sensor
    D. A knock sensor

18. With the engine running, the most obvious sign of a bad U-joint is a loud and hard clunk when:

    A.   coasting with the transmission in neutral.

    B.   placing the vehicle in DRIVE at a standstill.

    C.   trailing the throttle from a medium road speed.

    D.   shifting to neutral while at a standstill.

19. Technician A says that for cars and vans, staying within a 3% diameter change of the OEM tie size is acceptable. Technician B says that for pickups and sport utility vehicles (SUVs), room for up to a 15% oversize tire is generally provided for when the vehicle is designed. Who is correct?

    A.   A only

    B.   B only

    C.   Both A and B

    D.   Neither A nor B

20. The power steering fluid level is to be checked on a vehicle. Technician A says that on some cars the power steering fluid level can only be checked accurately after the engine has run for a brief period. Technician B says that on some cars a dipstick is used for checking the PS fluid only when it is "cold." Who is correct?

    A.   A only

    B.   B only

    C.   Both A and B

    D.   Neither A nor B

21. Abnormal noises made by a vane-type power steering pump may include any of the following EXCEPT a:

    A.   low frequency shudder.

    B.   low frequency moan.

    C.   high frequency whine.

    D.   high frequency hiss.

22. Which of the following is the correct procedure for adjusting toe on a vehicle?

    A.   Loosen and turn the adjusting sleeves with the proper tie rod sleeve tool.

    B.   Loosen and turn the castellated nuts on the tie rod ends.

    C.   Rotate the left tie rod end until the correct toe is achieved.

    D.   Rotate the right tie rod end until the correct toe is achieved.

23. If the part on an import vehicle as illustrated above is found to be leaking, which of the following describes how vehicle handling would be affected?

    A. The vehicle will continue to bounce up and down after hitting bumps.

    B. The vehicle will not track straight when the steering wheel is not held.

    C. The vehicle will lean excessively during turns.

    D. The vehicle will experience steering wheel jerkiness when hitting bumps.

24. Upon inspection, a vehicle's front ride height is low and the front coil springs show signs of coil contact. Which of the following should be done first?

    A. Replace the jounce bumpers.

    B. Replace the shocks.

    C. Replace the springs.

    D. Replace the tires.

25. A non-power steering vehicle with McPherson strut suspension is very difficult to turn to the left or right. Which of the following could be the cause?

    A. Faulty strut bearings

    B. Worn steering damper

    C. Worn struts

    D. Faulty tie rods

26. A rear leaf spring equipped vehicle seems to sag on the right rear side. Technician A says that a leaf in the left side leaf spring may have snapped. Technician B says that the right side spring shackle may be damaged. Who is correct?

    A. A only
    B. B only
    C. Both A and B
    D. Neither A nor B

27. A front wheel drive van with non-independent rear suspension shows severe rust and corrosion of the "U" shaped rear axle. Technician A says that such corrosion may weaken the axle and make it vulnerable to torsional stress. Technician B says that such a condition could lead to cracks or complete breakage of the axle. Who is correct?

    A. A only
    B. B only
    C. Both A and B
    D. Neither A nor B

28. The spindle assembly at the rear of the vehicle shows signs of severe wear. Which of the following could be the cause?

    A. Dry wheel bearings
    B. Incorrect rear camber
    C. Incorrect toe setting
    D. Worn rear tires

29. A growling noise is heard at the rear of an FWD independently suspended rear-end vehicle. Technician A says it could be caused by a bent spindle. Technician B says it could be caused by a faulty wheel bearing. Who is correct?

    A. A only
    B. B only
    C. Both A and B
    D. Neither A nor B

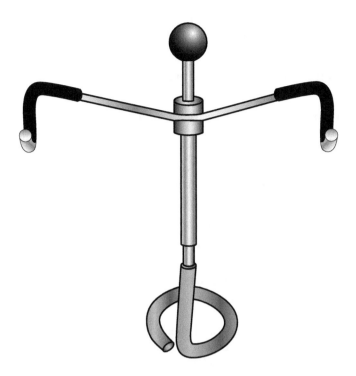

30. The object shown in the illustration above is used to:

    A. secure a steering wheel in position during a wheel alignment.

    B. hold the steering wheel during brake service.

    C. support a brake caliper once it has been unbolted from a wheel assembly.

    D. hold a seat assembly off the floor after it has been removed from a vehicle.

31. A new set of tires is to be installed on a vehicle. Technician A says that lower speed rated tires may be installed if the driver never drives the vehicle off his privately owned farm. Technician B says that the inflation pressure indicated on the tire sidewall is the inflation pressure which should be used. Who is correct?

    A. A only

    B. B only

    C. Both A and B

    D. Neither A nor B

32. Technician A says a tire puncture should be repaired before the tire assembly is balanced. Technician B says a tire may not need to be rebalanced if only a plug-type tire repair is performed and the wheel weights remained on the rim. Who is correct?

    A. A only

    B. B only

    C. Both A and B

    D. Neither A nor B

33. Following replacement of brake pads on a four-wheel disc brake equipped vehicle, the master cylinder reservoir is found to be overfilled and brake fluid is dripping from it. Which of the following is the most likely cause?

    A. The master cylinder port is clogged and causing brake fluid to back up.
    B. The reservoir is cracked and leaking brake fluid.
    C. The reservoir was not siphoned before the pistons were retracted.
    D. The brake calipers were not bled properly.

34. Following disassembly of a rear-wheel drum brake disassembly, the drums are found to be badly scored with blue areas on the braking surface. The technician should do all of the following EXCEPT:

    A. inspect the shoes to see if the rivets have contacted and scored the drums.
    B. measure the drums with a Vernier caliper to see if they may be reconditioned on a brake lathe and reused.
    C. inspect the wheel cylinders for brake fluid leakage.
    D. lubricate all self-adjusting hardware with high temperature waterproof grease.

35. Brake drums are to be turned using an off-car brake lathe. Technician A says to keep the lathe speed low when machining shallow cuts on the drums. Technician B says a rubber belt may be used around the circumference of the drum to keep it from getting hot while being machined. Who is correct?

    A. A only
    B. B only
    C. Both A and B
    D. Neither A nor B

36. The primary (smaller) shoe on a duo-drum brake system should be installed:

    A. before the secondary shoe.
    B. after the secondary shoe.
    C. toward the front of the vehicle.
    D. toward the rear of the vehicle.

37. After installation of a vehicle's wheels while on a lift, all of the wheel lug nuts should be:

    A. tightened by hand before the vehicle is lowered to the ground.
    B. tightened lightly with a wrench before the vehicle is lowered to the ground then fully torqued.
    C. fully tightened with an impact wrench before the vehicle is lowered to the ground.
    D. fully torqued to specs before the vehicle is lowered to the ground.

38. While inspecting the brakes on a disc-brake equipped vehicle, a maintenance technician finds that the brake caliper slides are severely corroded and dry of lubricant. In most cases, the correct action would be to:

    A. disregard their condition.
    B. replace the calipers.
    C. disassemble the brake assemblies; clean and lubricate the slides.
    D. replace the corroded slides with new ones.

39. A technician finds that a brake caliper has stripped mounting threads. Which of the following should be done?

    A. Use thread restorer on the damaged threads.

    B. Use a tap to restore the threads.

    C. Tear down the caliper and replace the defective part.

    D. Replace the caliper.

40. Rotors are being turned using an on-car brake lathe. Technician A says that using an on-car lathe is quicker than using an off-car brake lathe. Technician B says that some on-car brake lathes use the vehicle's engine to rotate the brake rotor while it is being turned on the vehicle. Who is correct?

    A. A only

    B. B only

    C. Both A and B

    D. Neither A nor B

41. The purpose of burnishing brake pads is to:

    A. seat the pads in the rotor grooves.

    B. transfer pad material to the rotors.

    C. make certain the brakes are not spongy.

    D. eliminate air from the brake hydraulic system.

42. A vacuum booster fails to fully assist the power brakes on a vehicle. Technician A says that the vacuum source to the booster should be checked with a vacuum gauge. Technician B says that the vacuum check valve should be checked for leakage. Who is correct?

    A. A only

    B. B only

    C. Both A and B

    D. Neither A nor B

43. Technician A says that with some electro-hydraulic braking systems, the high-pressure reservoir supplies the required brake pressure quickly and precisely to the wheel brakes without driver involvement. Technician B says that some electro-hydraulic braking systems offer improved active safety when braking in a corner or on a slippery surface. Who is correct?

    A. A only

    B. B only

    C. Both A and B

    D. Neither A nor B

44. While checking for voltage drop in a charging system while the charging system is operating, a technician finds 0.9 volts being dropped between the alternator and the battery on the insulated side. Which of the following is the most likely cause?

    A. A faulty voltage regulator is causing the voltage drop.

    B. The alternator's B+ ring terminal connection has burned where the cable is attached.

    C. The battery ground cable is corroded.

    D. This is an acceptable amount of voltage drop.

45. Which of the following could cause unwanted resistance in a tail light circuit?

    A. A larger-than-specification tail lamp

    B. The addition of a trailer harness

    C. The addition of an additional stop/tail light on a trailer hitch receiver

    D. Water leakage into a tail light housing

46. A standard flooded lead-acid battery is being replaced with an absorbed glass mat (AGM) battery. Technician A says that an AGM battery is less rugged than a flooded cell battery. Technician B says that under normal conditions, an AGM battery does not emit harmful hydrogen gas. Who is correct?

    A. A only

    B. B only

    C. Both A and B

    D. Neither A nor B

47. Which of the following is best suited for fully charging a completely discharged 12-volt battery?

    A. Charge the battery for at least 30 minutes by racing the vehicle's engine.

    B. Charge the battery for at least 1 hour by idling the engine.

    C. Remove the battery from the vehicle and trickle charge it for 6 hours.

    D. Remove the battery from the vehicle and slow charge it overnight.

48. Which of the following is LEAST LIKELY to cause excessive starter current draw?

    A. A partially seized engine

    B. A shorted field winding

    C. Faulty armature brushes

    D. A shorted armature

49. Before a starter is removed from an engine, it's important to:

    A. spot the engine to cylinder #1 TDC.

    B. spot the engine to cylinder #1 BDC.

    C. disconnect the positive lead from the battery.

    D. disconnect the negative lead from the battery.

50. The right side headlight fails a safety inspection because it does not work. Which of the following is the LEAST LIKELY cause?

    A. The bulb is burned out.

    B. The socket is corroded.

    C. The right side fuse has blown.

    D. The bulb is the incorrect type.

51. A horn fails to operate as it should. Technician A says the horn relay may be stuck closed. Technician B says the horn button may be stuck closed. Who is correct?

    A. A only
    B. B only
    C. Both A and B
    D. Neither A nor B

52. An A/C clutch fails to operate when the A/C is turned to ON. Any of the following could be the cause EXCEPT:

    A. the A/C system is low on refrigerant.
    B. the A/C clutch fuse has blown.
    C. the A/C clutch has a shorted winding.
    D. the A/C cooling fan fails to operate.

53. Which of the following is supposed to cool and return refrigerant from a gaseous state to a liquid state?

    A. The condenser
    B. The A/C compressor
    C. The evaporator
    D. The receiver-drier

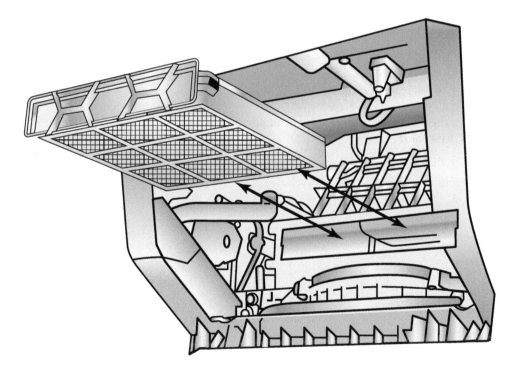

54. Which of the following parts is being replaced in the illustration above?

    A. The cabin air filter
    B. The A/C evaporator
    C. The intake air filter
    D. The glove box

55. The A/C serpentine drive belt keeps jumping off its pulley whenever the A/C is turned on. Which of the following is the most likely cause?

    A. A pulley or bracket is belt is misaligned.
    B. The belt is too small.
    C. The belt is made of the wrong material.
    D. The belt is too old.

PREPARATION EXAM 5

1. Before starting work on any vehicle, it's important to:

    A. gather as much information as you can.
    B. check with the customer.
    C. check with the shop supervisor.
    D. raise the vehicle on the lift.

2. While listening to a manual transmission vehicle engine as it is idling, the technician notes a loud ticking noise coming from the upper part of the engine. Which of the following is the most likely cause?

    A. A rod knock
    B. A worn thrust bearing
    C. A collapsed lifter
    D. A worn main bearing

3. All of the following cooling system components should be regularly inspected and tested EXCEPT for the:

    A. radiator.
    B. condenser.
    C. heater core.
    D. pressure cap.

4. What is the purpose of the tool illustrated above?

   A. To remove worm-type hose clamps
   B. To remove flat-band type hose clamps
   C. To remove wire-type hose clamps
   D. To remove double-wire-type hose clamps

5. Which of the following is commonly found on vehicles built through the 80s?

   A. Electric water pumps
   B. Belt-driven water pumps
   C. Vacuum-operated windshield wipers
   D. Draft tubes

6. Technician A says that after many miles, oil can accumulate in the induction system. Technician B says that carbon buildup in the throttle body can make the throttle blade stick closed. Who is correct?

   A. A only
   B. B only
   C. Both A and B
   D. Neither A nor B

7. Technician A says that one way to check for rusty exhaust pipes is to tap the pipes with ball peen hammer and listen for a metallic "ring." Technician B says that a missing heat shield from a catalytic converter could inadvertently start a fire. Who is correct?

    A. A only
    B. B only
    C. Both A and B
    D. Neither A nor B

8. The tool shown in the illustration above is used to disconnect:

    A. brake lines.
    B. heater hoses.
    C. fuel lines.
    D. EVAP hoses.

9. Diesel exhaust fluid (DTF) is to be added to a clean diesel vehicle. The cap on the DTF reservoir/bottle is normally:

    A. red.
    B. blue.
    C. white.
    D. yellow.

10. A test drive reveals that an electronically controlled automatic transmission equipped vehicle starts out in 2nd gear. Technician A says to check for diagnostic trouble codes (DTCs). Technician B says the transmission pan may have to be drained and removed to replace faulty transmission components. Who is correct?

    A. A only
    B. B only
    C. Both A and B
    D. Neither A nor B

11. Technician A says that automatic transmission fluids (ATFs) are supplied in many different colors. Technician B says that ATF can be smelled to determine if the clutches in the transmission are failing. Who is correct?

    A. A only
    B. B only
    C. Both A and B
    D. Neither A nor B

12. A CV joint boot is found to be dried and nearly split, but not yet leaking lubricant. The boot may be replaced by any of the following methods EXCEPT:

    A. replace the boot with a split-type replacement to avoid removing the axle-shaft.
    B. remove the axle-shaft and install a new boot.
    C. remove and replace the entire axle-shaft assembly to save time.
    D. remove, clean and repack the CV joint; install a new boot.

13. Upon inspection, a leak is found at the transmission cooler line fitting on a vehicle's radiator. Technician A says that tightening the threaded-type fitting to specs may cure the problem. Technician B says that replacing the rubber sealing ring may be all that is needed. Who is correct?

    A. A only
    B. B only
    C. Both A and B
    D. Neither A nor B

14. During a vehicle test drive, a driveline vibration is noticed. Technician A says that the driveshaft may be out of balance. Technician B says the driveshaft may be out of phase. Who is correct?

    A. A only
    B. B only
    C. Both A and B
    D. Neither A nor B

15. As shown in the illustration above, driveshaft angles (phasing) are being checked in a vehicle. If they are not both the same angle, which of the following might be likely to happen?

    A. The clutch might slip.
    B. The transmission might pop out of gear.
    C. The analog speedometer needle might fluctuate.
    D. The driveshaft might vibrate.

16. After a vehicle test drive, differential lubricant is seen leaking at the rear axle flange area and the axle tube is hot. Technician A says that the incorrect differential lubricant may have been used. Technician B says that the axle seal has failed and the axle bearing should be inspected for damage. Who is correct?

    A. A only
    B. B only
    C. Both A and B
    D. Neither A nor B

17. A broken wheel stud is to be replaced in an axle flange. Which of the following may be required to install the replacement?

    A. A large ball peen hammer
    B. A press
    C. A pair of slip-joint pliers
    D. A "C" clamp

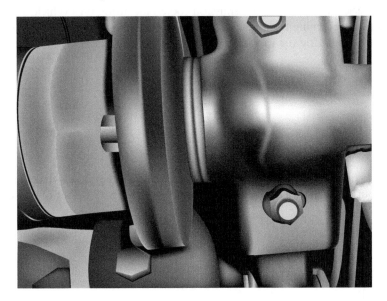

18. The front end of a 4WD vehicle is being inspected for leaks. Which of the following is likely to be causing the oil residue from a leak, as shown in the illustration above?

    A. A leaking universal joint
    B. A leaking gasket
    C. A leaking seal
    D. A leaking CV joint

19. An axle seal is found to be leaking on a 4WD vehicle. Which of the following is the most likely cause?

    A. Too little transfer case lubricant
    B. A plugged transaxle vent
    C. Driving over rough non-road surfaces
    D. Excessive turning on dry pavement while in 4WD

20. When a serpentine belt is replaced on a front-wheel drive vehicle, it's important to also make sure all of the following are done EXCEPT:

    A. make certain that the belt is the correct size and is properly seated in all of the grooved pulleys.

    B. check that the tensioner is free to flex and maintain tension on the belt.

    C. make certain belt dressing is applied to the belt to keep it running quietly and slip-free.

    D. make certain that the tensioner and pulleys do not make any noise as they operate.

21. After a power steering pressure hose has been replaced, it's important to:

    A. bleed the system of fluid.

    B. purge the system of air.

    C. drain and flush the system.

    D. pressurize the system.

22. An idler arm in a pickup truck is worn and needs to be replaced. Which of the following will be needed to complete the job?

    A. A ball joint separator

    B. A large ball peen hammer

    C. A wheel alignment

    D. New tie rod ends

23. Ride height is to be checked on a vehicle and compared to specs. Technician A says that ride height is the shortest distance between a level surface and the vehicle. Technician B says that the measurement should be taken with the vehicle loaded with average cargo and a driver. Who is correct?

    A. A only

    B. B only

    C. Both A and B

    D. Neither A nor B

24. Worn or damaged rebound bumpers often indicate that there is a need to replace the:

    A. springs.

    B. shock absorbers.

    C. upper control arms.

    D. lower control arms.

25. The front strut assemblies on a vehicle are being inspected. Technician A says to check that the struts are not bent, broken, or leaking oil. Technician B says when performing the "bounce test" the vehicle should not bounce up and down more than four times. Who is correct?

    A. A only

    B. B only

    C. Both A and B

    D. Neither A nor B

26. An older vehicle shows signs of dried and split jounce bumpers. Technician A says that this may cause the vehicle to bounce excessively after hitting a bump. Technician B says that the vehicle may experience unstable tracking. Who is correct?

   A. A only

   B. B only

   C. Both A and B

   D. Neither A nor B

27. With a vehicle's wheels of the ground, the rear struts are found to be loose at their upper end. Which of the following should be done first?

   A. Replace the struts.

   B. Inspect the upper mounting plates.

   C. Replace the mounting plates.

   D. Replace the strut bushings.

28. Which of the following would LEAST LIKELY be affected by loose rear tie rods on an independent rear-suspension vehicle?

   A. Toe setting

   B. Vehicle tracking

   C. Rear sway

   D. Ride comfort

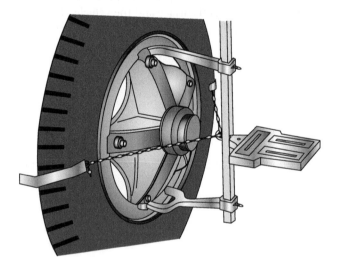

29. Which of the following is being measured using the setup in the illustration above?

   A. Included angle

   B. Camber

   C. Toe

   D. Steering Axis inclination

30. Technician A says most front-wheel drive vehicles use a degree of toe out to enable the front tires to run parallel to each other at road speeds. Technician B says that toe is one of the most critical alignment settings relative to tire wear. Who is correct?

   A. A only
   B. B only
   C. Both A and B
   D. Neither A nor B

31. When a tire is being mounted on a wheel rim, it's important to match mount the tire on the rim:

   A. to help minimize the final combination of force variation and/or imbalance.
   B. to avoid damage to the tire pressure monitoring system (TPMS) sending unit.
   C. so that the tire does not have to be rebalanced.
   D. to avoid damage to the tire.

32. A replacement tire pressure monitoring system (TPMS) sensor has been installed in a vehicle's left-rear tire and rim assembly. Technician A says to use a TPMS activator to perform a relearn procedure on the new sensor. Technician B says that the three other tires and the spare will also need a relearn procedure performed. Who is correct?

   A. A only
   B. B only
   C. Both A and B
   D. Neither A nor B

33. A metal brake line at the rear of a vehicle is found to be dented by a floor jack. Which of the following should be done?

   A. Replace the dented section using a stainless steel compression fitting.
   B. Replace the dented section using a brass compression fitting.
   C. Replace the dented section using a copper compression fitting.
   D. Replace the dented brake line from fitting to fitting with a steel brake line.

34. A brake warning light bulb fails to operate. Technician A says the warning brake light bulb may be faulty. Technician B says the park brake light switch may be misadjusted. Who is correct?

   A. A only
   B. B only
   C. Both A and B
   D. Neither A nor B

35. Which of the following should be used when lubricating brake shoe support pads?

   A. Vaseline®
   B. Chassis lubricant
   C. WD-40®
   D. Water/heat resistant grease

36. Technician A says a wheel cylinder can be rebuilt using a replacement cups and boots kit. Technician B says many wheel cylinder housings are made from aluminum making honing not possible, so the entire wheel cylinder assembly should be replaced. Who is correct?

    A. A only

    B. B only

    C. Both A and B

    D. Neither A nor B

37. Parking brake cables should be lubricated with:

    A. Permatex®.

    B. WD-40®.

    C. Wheel bearing grease.

    D. Graphite lubricant.

38. A disc brake caliper is being dismounted for brake pad replacement. Technician A says the caliper may be hung by the brake hose during the repair. Technician B says that an air gun may be used to remove brake dust from the caliper. Who is correct?

    A. A only

    B. B only

    C. Both A and B

    D. Neither A nor B

39. To remove a hubless brake rotor from a vehicle with disc brakes, the technician should first:

    A. unbolt the lug nuts, remove the wheel assembly and lift the rotor off the wheel studs.

    B. unbolt the lug nuts, remove the wheel assembly and remove the castellated nut and slide the rotor off the spindle with the outer bearing parts.

    C. unbolt the lug nuts, remove the wheel assembly and use a slide hammer to remove the hub assembly.

    D. unbolt the lug nuts, remove the wheel assembly and unbolt and swing the caliper aside.

40. A parking brake using an integral-style caliper is to be adjusted. Technician A says to depress the service brake pedal several times. Technician B says to check and, if necessary, to adjust the parking brake cable to the proper length. Who is correct?

    A. A only

    B. B only

    C. Both A and B

    D. Neither A nor B

41. Wheel lug nuts should be properly torqued to specs for all of the following reasons EXCEPT:

    A. to help center the wheel.

    B. to make sure the wheel does not come off.

    C. to prevent brake rotor warpage.

    D. To avoid wheel imbalance.

42. To test for brake pedal travel, push the brake pedal a few times to relieve vacuum in the brake booster, hold the brake pedal down firmly, and:

    A. turn on the ignition and note the change of brake pedal position.

    B. start the engine and note the amount of change to the brake pedal position.

    C. start and rev up the engine and note the change of brake pedal position.

    D. accelerate and hold the engine at 1500 rpm and note the brake position change.

43. With a vehicle idling, a loud hiss is heard and the engine stalls whenever the brake pedal is applied. Which of the following is the most likely cause?

    A. A faulty brake master cylinder

    B. A leaking vacuum line to the brake booster check valve

    C. A leaking brake booster

    D. A misadjusted brake pedal pushrod

44. When the ignition system of a vehicle is first turned on, the supplemental restraint system (SRS/airbag) light comes on and then goes off. This indicates that the SRS:

    A. lamp burned out.

    B. fuse is blown.

    C. has a fault and has stored a DTC.

    D. self-test has been successfully completed.

45. Which of the following would be LEAST LIKELY to cause keep alive memory (KAM) of engine parameters to be lost?

    A. A blown fuse for the EBCM (ABS) module

    B. A burned fusible link

    C. A blown fuse for the powertrain control module (PCM)

    D. A dead battery

46. The B+ battery cable connection at the battery appears to be clean, yet a voltage drop test reveals a 1.6 volt drop in the B+ cable to the starter when the engine is being cranked. Technician A says the B+ cable may be corroded under its insulation. Technician B says the B+ connection at the starter may be oil soaked or loose. Who is correct?

    A. A only

    B. B only

    C. Both A and B

    D. Neither A nor B

47. Which of the following best describes the last connection to be made while jump starting a vehicle with a dead battery using another vehicle:

    A. Connect the positive to the positive battery terminal of vehicle to be jump started.

    B. Connect the positive to the negative battery terminal of a vehicle to be jump started.

    C. Connect the positive to the engine block of the vehicle to be jump started.

    D. Connect the negative to the engine block of the vehicle to be jump started.

48. A neutral safety switch is suspected of being faulty. Technician A says that an easy way to bypass the suspected switch would be to use a jumper wire across the starter relay terminals at the fuse panel. Technician B says the entire starter control circuit can be bypassed by using a remote starter switch at the starter solenoid. Who is correct?

    A. A only
    B. B only
    C. Both A and B
    D. Neither A nor B

49. A vehicle's alternator is to be removed from a vehicle for testing. The first step is to remove the:

    A. two or three mounting bolts holding the alternator to the bracket.
    B. alternator bracket.
    C. battery positive cable connector from the alternator.
    D. battery negative cable connector at the battery.

50. A vehicle's headlight lens is fogged over with moisture on the inside. Which of the following is the LEAST LIKELY cause?

    A. A small stone-impact hole in the lens
    B. A crack at the base of the headlight bulb
    C. A missing "O" ring at the headlight socket
    D. A faulty gasket around the perimeter of the lens

51. A front wiper blade's rubber is found to be torn and hanging from the wiper blade. Technician A says that just the rubber insert can be replaced. Technician B says it's faster and easier to replace the entire wiper blade. Who is correct?

    A. A only
    B. B only
    C. Both A and B
    D. Neither A nor B

52. Whenever a vehicle is heavily accelerated, air blowing from the dashboard vents shifts to the defroster vents. Which of the following is the most likely cause?

    A. A defective blower motor
    B. A faulty vacuum check valve
    C. A faulty blower switch
    D. A faulty blend door sensor

53. A leak is detected at the lower left corner of an A/C condenser. Technician A says the leak can be fixed by soldering it. Technician B says the leak can be fixed by using an epoxy sealant. Who is correct?

    A. A only
    B. B only
    C. Both A and B
    D. Neither A nor B

54. If a cabin air filter is to be replaced, it may be found located either under the dashboard or:

A. under the vehicle.

B. under the passenger's seat.

C. inside the engine compartment.

D. in the headliner above the passenger's seat.

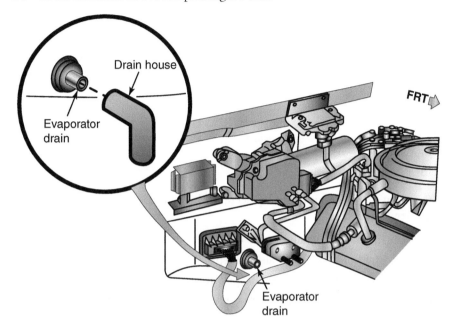

55. The drain hose illustrated above, drips water when the A/C is ON, but not at other times. Which of the following is the most likely cause?

A. Condensate is draining from the evaporator.

B. Water is leaking from the evaporator.

C. There is a leak in the heater core.

D. The condenser is leaking.

PREPARATION EXAM 6

1. A vehicle's exhaust emits blue smoke. Which of the following is the most likely cause?

A. A dirty air filter

B. A stuck-open crankcase ventilation system

C. A faulty spark plug

D. Leaking valve stem seals

2. Gasoline is leaking from an older vehicle's metal fuel tank. Technician A says the fuel tank may have been over-filled. Technician B says the fuel tank may have rusted from water lying at the bottom of the tank. Who is correct?

    A. A only

    B. B only

    C. Both A and B

    D. Neither A nor B

3. Technician A says that some coolant recovery tanks should not be opened until cooled off. Technician B says that a pressure test would confirm if a radiator cap cannot hold pressure. Who is correct?

    A. A only

    B. B only

    C. Both A and B

    D. Neither A nor B

4. A cooling system thermostat is being replaced. Which of the following should be done when installing a new one?

    A. The gasket should be liberally coated with RTV sealant.

    B. The thermostat housing should also be replaced.

    C. The small air bleed hole should be located at the top.

    D. The thermostat should be adjusted to open at 180 degrees Fahrenheit.

5. A timing belt with 100,000 miles on it is being replaced. Technician A says to replace the water pump at the same time. Technician B says to replace the tensioner at the same time. Who is correct?

    A. A only

    B. B only

    C. Both A and B

    D. Neither A nor B

6. Whenever the weather is rainy, an engine's throttle blade sticks closed, making it difficult to accelerate off idle. Which of the following is the most likely cause?

    A. Carbon is deposited on the throttle body throat and the throttle blade.

    B. The throttle cable is sticking.

    C. The throttle cable pulley is warped.

    D. The cruise control cable is binding the throttle cable.

7. A PCV valve does not rattle when it is shaken. Which of the following would be best to do?

    A. Soak it in a parts cleaning tank.

    B. Spray it with carburetor or fuel injector cleaner.

    C. Replace it.

    D. Use it as it is; there is nothing wrong with the valve.

8. Upon removal and inspection of an engine's spark plugs, oily deposits are found on the insulator of one spark plug. Which of the following is the most likely cause?

   A. The spark plug is faulty.

   B. A valve stem seal is leaking.

   C. The piston rings are worn.

   D. A spark plug coil wire is faulty.

9. Which of the following names is commonly used when discussing diesel exhaust fluid (DEF)?

   A. Adware

   B. Amsol

   C. AdBlut

   D. AdBlue

10. During a test drive, an older model automatic transmission vehicle is found to not shift properly. Any of the following could be the cause EXCEPT:

   A. low transmission fluid.

   B. too much transmission fluid.

   C. an inoperative torque converter clutch.

   D. worn plastic teeth on a governor drive gear.

11. Which of the following would cause an automatic transmission to leak fluid from the bell housing?

   A. A faulty pan gasket

   B. A faulty front pump seal

   C. A faulty rear seal

   D. A leaking seal at the shift linkage

12. Technician A says a transmission cooler may be located under the car. Technician B says an additional transmission cooler may be installed behind the radiator. Who is correct?

   A. A only

   B. B only

   C. Both A and B

   D. Neither A nor B

13. Worn engine mounts are to be replaced. All of the following steps may need to be done EXCEPT:

   A. raise the engine with an engine hoist.

   B. disconnect the upper ball joints.

   C. remove the lower cross-member.

   D. drop the power steering rack from its mountings.

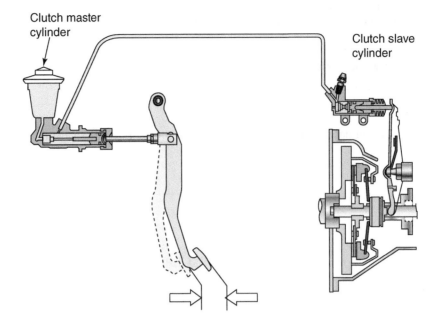

Clutch master cylinder

Clutch slave cylinder

14. At which of the locations illustrated above is air to be bled from the hydraulic clutch system?

   A. From the clutch slave cylinder

   B. From the fluid reservoir

   C. From the clutch master cylinder

   D. From the master cylinder

15. All of the following are used as either manual transmission or transaxle gear lubricants EXCEPT:

   A. 30 weight motor oil.

   B. 80W90 weight gear oil.

   C. Hypoid gear oil.

   D. P.S. fluid.

16. A solid rear-axle is leaking lubricant past its seal onto drum-style wheel brakes. Technician A says the axle will need to be removed to replace the seal. Technician B says the brakes at both rear wheels will need to be replaced. Who is correct?

   A. A only

   B. B only

   C. Both A and B

   D. Neither A nor B

17. If the differential vent on a solid rear-axle vehicle becomes clogged, which of the following would likely happen first?

   A. The axle bearings could become overheated.

   B. The axle bearings could fail.

   C. The axle seal(s) could leak differential fluid.

   D. The axle bearings could make noise.

18. The right front CV joint on a 4WD vehicle "clicks" loudly during sharp turns. Technician A says the CV joint needs to be replaced. Technician B says that the right front wheel may ultimately quit steering the vehicle. Who is correct?

    A. A only

    B. B only

    C. Both A and B

    D. Neither A nor B

EXAMPLE: P0137 LOW VOLTAGE BANK 1 SENSOR 2

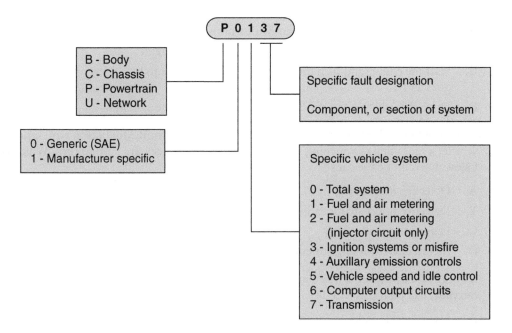

19. A failure has occurred in a 4WD vehicle's transmission control module (TCM). Based upon the chart above, which of the following diagnostic trouble codes (DTCs) would be stored?

    A. B0348

    B. C0045

    C. P0700

    D. U0005

20. Following replacement of a power steering pump, as well as its pulley, the serpentine drive belt keeps jumping off. Upon investigation, the pulley is noted to be out of alignment by 1/4 inch. Which of the following is the most likely cause?

    A. The pulley is defective.

    B. The power steering pump splines are nicked.

    C. The power steering bracket is incorrect.

    D. The pulley was not pressed all the way onto the PS pump shaft.

21. Following the replacement of a power steering (PS) pressure hose, the steering system makes a groaning/growling noise when the steering wheel is turned. Technician A says an incorrect PS pressure hose was installed. Technician B says there is air in the PS system. Who is correct?

    A. A only

    B. B only

    C. Both A and B

    D. Neither A nor B

22. Which of the following is the proper method of purging air from a power steering (PS) system?

    A. Prime the power steering system with a pressure bleeder.

    B. Open the power steering bleed screw and start the engine.

    C. Start the engine and turn the steering wheel from lock to lock a few times until the noise goes away.

    D. Drive the vehicle a few miles to get rid of the air in the system.

23. When the front wheels are turned in either direction on a front McPherson strut equipped vehicle, a squeaking/scrunching noise is heard, and the wheels are hard to turn. Which of the following could be the cause?

    A. Defective front struts

    B. Defective upper strut bearings

    C. Faulty lower ball joints

    D. Incorrect camber adjustment

24. All of the following are a true statements about replacing a vehicle's upper (non-load carrying) ball joints EXCEPT:

    A. Use a "pickle fork" to loosen the worn ball joints.

    B. Compress the coil springs before replacing the ball joints.

    C. Re-align the front end after replacing the ball joints.

    D. Install a "Zerk" fitting on the ball joint after installation.

25. A front steering knuckle is being inspected for damage after a vehicle accident. Technician A says that because of the steering knuckles design, a bent condition may not be easy to see. Technician B says that if a steering knuckle is bent, it might only be apparent when a wheel alignment is attempted. Who is correct?

    A. A only

    B. B only

    C. Both A and B

    D. Neither A nor B

26. Sway bar bushings are to be replaced. Technician A says that worn sway bar bushings can affect vehicle tracking. Technician B says that following sway bar bushing replacement, a wheel alignment will be needed. Who is correct?

    A. A only

    B. B only

    C. Both A and B

    D. Neither A nor B

27. A front wheel bearing is failing due to hard use. Which of the following is the LEAST LIKELY symptom?

    A. A whining noise will be heard.
    B. A growling noise will be heard on turns.
    C. A noise similar to driving on snow tires may be heard.
    D. The front wheel assembly may experience play when grasped and moved.

28. A vehicle with a solid rear axle has a broken panhard rod (tracking bar). Which of the following will the vehicle likely experience?

    A. Rear end lateral movement or sway
    B. Poor tire life
    C. Excessive bouncing
    D. Drifting at higher speeds

29. The grease fitting on a rear ball joint has receded into the ball joint. Technician A says that the wear-indicator type ball joint should be replaced. Technician B says that ball joints on both sides should be replaced even if one side only is worn. Who is correct?

    A. A only
    B. B only
    C. Both A and B
    D. Neither A nor B

30. Shock absorber mounts can be checked for good condition by using any of the following methods EXCEPT:

    A. inspecting them for dryness and cracking.
    B. observing them while bouncing the vehicle.
    C. performing a wheel alignment.
    D. watch the mounts while grasping the shock firmly and shaking it.

31. The front caster setting on a McPherson strut equipped vehicle is found to be incorrect. Technician A says that there may be no OEM provision for making a caster adjustment. Technician B says that aftermarket parts may enable caster to be adjusted on a front-strut equipped vehicle. Who is correct?

    A. A only
    B. B only
    C. Both A and B
    D. Neither A nor B

32. Technician A says that tire wear at the outside edge of a tire could be caused by too much negative camber. Technician B says that too much wear on both outer edges of a tire could be caused by over-inflation of the tire. Who is correct?

    A. A only
    B. B only
    C. Both A and B
    D. Neither A nor B

33. Technician A says that contaminated brake fluid could cause a spongy brake pedal.
     Technician B says that contaminated brake fluid could cause corrosion of the brake pads.
     Who is correct?

     A. A only
     B. B only
     C. Both A and B
     D. Neither A nor B

34. The most commonly used brake fluid is:

     A. DOT 3.
     B. DOT 4.
     C. DOT 5.
     D. silicone brake fluid.

35. Technician A says that a brake drum may be machined until the dimension shown in the
     illustration above is reached. Technician B says that the drum should be inspected for hard
     spots and blue areas from excessive heat. Who is correct?

     A. A only
     B. B only
     C. Both A and B
     D. Neither A nor B

36. The primary lubrication points of drum-type wheel brake assemblies include all of the
     following EXCEPT:

     A. the raised pads on the backing plates.
     B. the interior surface of the brake drum.
     C. the contact points of self-adjusters.
     D. the star wheel brake adjuster.

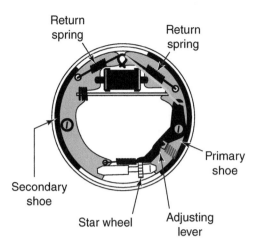

37. Referring to the illustration above, at which location are the brakes initially adjusted by hand for proper brake operation?

   A.   The primary shoe
   B.   The secondary shoe
   C.   The star wheel
   D.   The adjusting lever

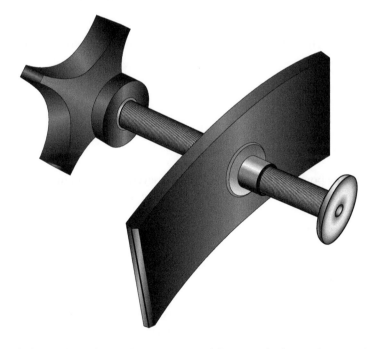

38. A disc brake caliper is being removed from a vehicle. Technician A says the tool shown above, may be used to retract the caliper piston. Technician B says the tool shown above, may be used to remove and install the brake pads. Who is correct?

   A.   A only
   B.   B only
   C.   Both A and B
   D.   Neither A nor B

39. Brake rotor runout can best be checked by using a(an):

    A. inside micrometer.

    B. dial indicator.

    C. snap gauge.

    D. feeler gauge.

40. Which of the following is LEAST LIKELY to help prevent brake pads from making a squealing noise?

    A. Brake pad paste

    B. Mechanic silencers

    C. Bearing grease

    D. Water- and heat-resistant lubricant

41. Technician A says that on some vehicles pushing a button on the dashboard in the correct sequence will turn off the Brakes light. Technician B says that after fluid is added to a brake master cylinder, the "Brake Warning" light should go off. Who is correct?

    A. A only

    B. B only

    C. Both A and B

    D. Neither A nor B

42. Burnishing new brake pads may require aggressively stopping a vehicle from 30 mph, at 30 second intervals, as many as:

    A. 10 times.

    B. 20 times.

    C. 30 times.

    D. 40 times.

43. The vacuum supplied to the power brake booster leaks out several minutes after the engine is shut off. Any of the following could be the cause EXCEPT for a leaking:

    A. brake booster.

    B. vacuum check valve.

    C. vacuum hose upstream of the check valve.

    D. vacuum hose downstream of the check valve.

44. A vehicle experiences the MIL light turning on while driving. After connecting a scan tool, the technician finds several diagnostic trouble codes (DTCs) displayed at the same time. The most likely cause is:

    A. an incorrect voltage supply.

    B. an open circuit.

    C. a short circuit.

    D. a corroded ground connection.

45. A fully charged 12-volt flooded cell automotive battery should have an open circuit voltage of

A. 10.2 volts.

B. 12.0 volts.

C. 12.6 volts.

D. 13.2 volts.

46. A flooded lead acid battery with removable caps should be only be topped off with:

A. sulfuric acid.

B. tap water.

C. carbonated water.

D. distilled water.

47. A technician is not able to hear a manual transmission vehicle "click," or crank, when the ignition key is moved to the START position. Which of the following could be the most likely cause?

A. A low battery

B. Dirty battery connections

C. A faulty neutral safety switch

D. The clutch pedal not fully depressed

48. To test the charging system in the field, it's OK to:

A. remove the B+ battery cable from the battery while the engine is running to see if it stays running.

B. rev the engine and see if a test light glows brighter.

C. connect an ammeter across the battery and measure charging amp.

D. check battery open circuit voltage when the engine is OFF, and compare it to when the engine is running at 2000 rpm.

49. Fog lights fail to operate when the high beams are on but work OK with the low beams. Which of the following is the most likely cause?

A. The high beam switch is defective.

B. The fog light switch is faulty.

C. The fog light fuse is blown.

D. Fog lights are not supposed to work with the high beams on.

50. Technician A says a scan tool can be used to reset some vehicles' "Change Oil Soon" maintenance reminder lights. Technician B says that the menu for the on-board Driver Information Center used on some vehicles can be used to reset the oil change reminder. Who is correct?

A. A only

B. B only

C. Both A and B

D. Neither A nor B

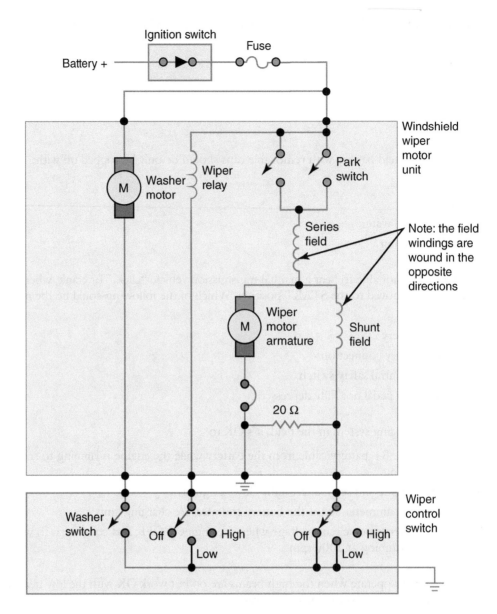

51. Referring to the schematic diagram above, Technician A says that the washer motor can be tested by connecting a wire from the battery to the wiper relay input. Technician B says to ground the input terminal at the washer switch to make the washer motor operate. Who is correct?

A. A only

B. B only

C. Both A and B

D. Neither A nor B

52. A leak detector is being used to check an A/C system. Technician A says to move the wand slowly along the top of all fittings, hoses, etc. Technician B says to use an open flame to check for leaks. Who is correct?

    A.   A only
    B.   B only
    C.   Both A and B
    D.   Neither A nor B

53. A cabin air filter is found to be clogged with debris. Technician A says that this could be the result of leaves getting into the cowling. Technician B says this could be caused by improper car washing. Who is correct?

    A.   A only
    B.   B only
    C.   Both A and B
    D.   Neither A nor B

54. An A/C "V" drive belt is being adjusted. The proper tension can be determined by using a:

    A.   ruler.
    B.   scale.
    C.   Vernier caliper.
    D.   protractor.

55. A good way to clean an evaporator drain is to use:

    A.   a stiff brush.
    B.   a coat hanger.
    C.   a shop vacuum.
    D.   compressed shop air.

6 Answer Keys and Explanations

INTRODUCTION

Included in this section are the answer keys for each preparation exam, followed by individual, detailed answer explanations and a reference identifying the designated task area being assessed by each specific question. This additional reference information may prove useful if you need to refer back to the task list located in section 4 of this book for additional support.

PREPARATION EXAM 1—ANSWER KEY

1.	D	20.	A	39.	C
2.	D	21.	B	40.	B
3.	B	22.	B	41.	C
4.	D	23.	D	42.	D
5.	D	24.	A	43.	B
6.	C	25.	A	44.	D
7.	B	26.	C	45.	D
8.	C	27.	A	46.	B
9.	D	28.	B	47.	A
10.	C	29.	B	48.	D
11.	C	30.	A	49.	C
12.	D	31.	D	50.	C
13.	C	32.	B	51.	D
14.	D	33.	C	52.	B
15.	B	34.	D	53.	B
16.	D	35.	A	54.	B
17.	B	36.	D	55.	B
18.	B	37.	A		
19.	D	38.	B		

PREPARATION EXAM 1—EXPLANATIONS

1. While road testing a vehicle, a technician verifies the customer's complaint of steering wheel shimmy during hard braking. The technician should:

    A. tell the vehicle owner.
    B. tell the shop supervisor.
    C. tell the shop manager.
    D. note the concern on the repair order.

    TASK A.1

    Answer A is incorrect. The vehicle owner already knows about the concern.

    Answer B is incorrect. The shop supervisor does not need to know this at this time.

    Answer C is incorrect. The shop manager does not need to know this at this time.

    Answer D is correct. The fault should be noted on the repair order as something to be addressed.

2. While performing a scheduled maintenance test drive, which of the following noises would LEAST LIKELY be scheduled for repair?

    A. Belt squeal
    B. Brake squeal
    C. Tire noise
    D. Wind noise

    TASK A.3

    Answer A is incorrect. A belt squeal may mean the belt is faulty or needs adjusting.

    Answer B is incorrect. Brake squeal may mean worn shoes or pads.

    Answer C is incorrect. Tire noise may mean one or more faulty tires or an alignment issue.

    Answer D is correct. While important, wind noise is an issue that is not normally associated with a scheduled maintenance.

TASK A.7

3. Which type of replacement drive belt is shown in the illustration above?

A. A "V" belt

B. A serpentine belt

C. A toothed timing belt

D. A toothed "start/stop" drive belt

Answer A is incorrect. A serpentine belt is shown.

Answer B is correct. A serpentine belt is shown.

Answer C is incorrect. A timing belt is not shown.

Answer D is incorrect. A hybrid vehicle stop/start belt is not shown.

TASK A.8

4. It's a good idea to routinely replace cooling system hoses:

A. after 25,000 miles.

B. after 50,000 miles.

C. after 100,000 miles.

D. based on the OEM schedule.

Answer A is incorrect. Hoses may last far longer.

Answer B is incorrect. Some OEMs may consider this too soon; however, this is a good time to inspect the hoses.

Answer C is incorrect. Different OEMs recommend replacement at varying intervals; however, this could be a good time to change them

Answer D is correct. Always follow the OEM recommendations for hose replacement.

5. Engine coolant can be tested by using any of the following EXCEPT a(an):

   A. hydrometer.
   B. voltmeter.
   C. refractometer.
   D. ammeter.

TASK A.10

Answer A is incorrect. A hydrometer checks the specific gravity of coolant as an indication of freeze point.

Answer B is incorrect. A voltmeter can detect electrolysis caused by impurities in the coolant.

Answer C is incorrect. A refractometer checks the composition of the coolant.

Answer D is correct. An ammeter is not a useful tool for checking coolant.

6. Which of the following should illuminate if the charging system is not working?

   A. The ABS light
   B. The TPMS light
   C. The alternator light
   D. The check engine light

TASK A.13

Answer A is incorrect. The ABS light alerts the driver of anti-lock brake issues.

Answer B is incorrect. The TPMS light indicates low tire pressure(s).

Answer C is correct. The light comes on if alternator output drops below battery voltage.

Answer D is incorrect. The "check engine" light indicates an engine or emissions related issue.

7. A vehicle's air filter is found to be contaminated with engine oil. Which of the following is the most likely cause?

   A. A leaking PCV hose
   B. Excessive blowby
   C. A stuck closed PCV valve
   D. A faulty idle-air control motor

TASK A.16

Answer A is incorrect. A leaking PVC hose would not cause this problem.

Answer B is correct. Excessive blowby would force oil fumes up into the air cleaner housing.

Answer C is incorrect. A stuck-closed PCV valve would cut off crankcase gases from reaching the air filter.

Answer D is incorrect. The IAC controls idle air around the closed throttle plate, not crankcase emissions.

8. When installing replacement spark plugs, some manufacturers recommend coating the threads with:

   A. chassis lube.
   B. motor oil.
   C. anti-seize.
   D. WD-40®

TASK A.19

Answer A is incorrect. Chassis lubricant is not the correct lubricant.

Answer B is incorrect. Motor oil is not the correct lubricant.

Answer C is correct. If called for by the OEM, anti-seize will not burn onto and coke the spark plug threads.

Answer D is incorrect. WD-40® is not recommended to be used on spark plug threads as a lubricant.

TASK A.21

9. A pre-formed plastic pipe for the EVAP system is found to be cracked. Technician A says to wrap electrical tape around the crack to repair the leak. Technician B says to apply non-hardening sealant to the pipe. Who is correct?

A. A only

B. B only

C. Both A and B

D. Neither A nor B

Answer A is incorrect. This is a temporary and unreliable repair.

Answer B is incorrect. This too is an unreliable repair.

Answer C is incorrect. Both Technicians A and B are incorrect.

Answer D is correct. Neither Technician is correct. Neither tape nor sealant is to be used; the pipe needs to be replaced.

TASK B.3

10. An automatic transmission pan has been removed and is being reinstalled. Technician A says that some OEMs recommend using ATV sealant instead of a gasket. Technician B says that some OEMs recommend reusing the original gasket. Who is correct?

A. A only

B. B only

C. Both A and B

D. Neither A nor B

Answer A is incorrect. Some OEMs do advocate using ATV sealant instead of a gasket.

Answer B is incorrect. Some OEMs offer reusable transmission pan gaskets.

Answer C is correct. Both Technicians are correct.

Answer D is incorrect. Both Technicians A and B are correct.

TASK B.4

11. A CV joint boot is found to have a small split. Which of the following should be done?

A. Seal it with ATV.

B. Seal it with Permatex®.

C. Replace the boot.

D. Replace the half-shaft assembly.

Answer A is incorrect. This is a temporary fix that may work, but it is not the recommended solution.

Answer B is incorrect. This may work for a short time at best but is a temporary fix.

Answer C is correct. This is the proper repair.

Answer D is incorrect. Replacing the entire shaft, while perhaps convenient, is not the needed repair.

12. A transmission cooler has developed a leak. Technician A says the cooler may be reliably repaired with epoxy. Technician B says the cooler may be reliably repaired with ATV sealant. Who is correct?

TASK B.5

   A.   A only
   B.   B only
   C.   Both A and B
   D.   Neither A nor B

Answer A is incorrect. This is an unreliable repair.

Answer B is incorrect. This is also an unreliable repair.

Answer C is incorrect. Neither Technician is making a correct statement.

Answer D is correct. Neither Technician is correct. The cooler should be replaced.

13. Technician A says that some automatic transmission filters are made of pleated paper. Technician B says that some transmission filters require removal of the transmission pan for replacement. Who is correct?

TASK B.7

   A.   A only
   B.   B only
   C.   Both A and B
   D.   Neither A nor B

Answer A is incorrect. Some transmission filters are made of pleated paper.

Answer B is incorrect. In any cases the pan must be removed.

Answer C is correct. Both Technicians are correct.

Answer D is incorrect. Neither is making incorrect statements.

14. An oil-soaked engine mount:

TASK C.2

   A.   is OK and should be let alone.
   B.   should be removed, cleaned and reinstalled.
   C.   should be cleaned with spray solvent.
   D.   should be replaced.

Answer A is incorrect. The engine mount should be replaced because the oil will have weakened it.

Answer B is incorrect. Cleaning the mount will not make it any stronger; the engine mount should be replaced.

Answer C is incorrect. Solvent will not strengthen an oil-soaked engine mount; it needs to be replaced.

Answer D is correct. The engine mount(s) should be replaced.

TASK C.4

15. A manual RWD transmission is leaking at the rear seal. Technician A says the transmission will have to be removed from the vehicle. Technician B says that the drive shaft will have to be removed from the vehicle. Who is correct?

A. A only

B. B only

C. Both A and B

D. Neither A nor B

Answer A is incorrect. Transmission removal may not be necessary.

Answer B is correct. Only Technician B is correct. The driveshaft (with the U-joint) must be removed in order to replace the rear seal.

Answer C is incorrect. Only Technician B is correct.

Answer D is incorrect. Only Technician A's statement is incorrect.

TASK C.8

16. A drive shaft center support bearing is found to be rusty. The bearing should be:

A. sprayed with WD-40®.

B. lubricated with 90 weight oil.

C. lubricated with wheel bearing grease.

D. removed and replaced.

Answer A is incorrect. WD-40® would not fix a rusty bearing.

Answer B is incorrect. Oil would not cure the problem.

Answer C is incorrect. Greasing the bearing would be a temporary fix at best.

Answer D is correct. The bearing has gone dry and is rusted; it should be replaced.

TASK C.11

17. When replacing the gear lube in a differential, be sure to do all of the following EXCEPT:

A. drain the old gear lube into a pan.

B. fill the differential until the new fluid is 1 inch below the fill hole.

C. dispose of the old gear lube using a licensed recycler.

D. use new gear lube as recommended by the OEM.

Answer A is incorrect. The old gear lube should be caught in a drain pan or equivalent.

Answer B is correct. The statement is incorrect. Gear lube should be filled up to the fill hole level.

Answer C is incorrect. Any used automotive type oil or fluid should be recycled.

Answer D is incorrect. OEM specs should be followed; there are different requirements for different vehicles.

TASK C.14

18. A vehicle owner complains that it is difficult to get a 4WD vehicle into and out of 4WD. Which of the following is the most likely cause?

A. The transfer case is overfilled with gear lube.

B. The shift linkage is damaged.

C. The universal joints need to be lubricated.

D. The drive shaft is bent.

Answer A is incorrect. This is not a likely cause of difficult shifting.

Answer B is correct. Damaged linkage could make shifting difficult.

Answer C is incorrect. U-joints would not likely affect manual shifting ability.

Answer D is incorrect. A bent driveshaft would cause vibration, but not difficult shifting.

19. After having a tire replaced on a vehicle, a customer complains that the ABS light stays ON. Which of the following is the most likely cause?

    TASK C.20

    A. The ABS light circuit is open.

    B. The tire has been installed backwards.

    C. The wheel was bent during tire mounting.

    D. The tire is the wrong size.

    Answer A is incorrect. This would cause the light to stay OFF, certainly not cause the light to turn ON.

    Answer B is incorrect. A backwards directional tire would not cause the ABS light to illuminate.

    Answer C is incorrect. A bent wheel would not likely affect the ABS.

    Answer D is correct. A tire with a different radius could constantly rotate at a different speed than the rest of the wheels and trip an ABS light ON.

20. Technician A says that Supplemental Restraint System (SRS) electrical connectors are often yellow in color. Technician B says that SRS circuits must be allowed to power down for 10 seconds. Who is correct?

    TASK D.1

    A. A only

    B. B only

    C. Both A and B

    D. Neither A nor B

    Answer A is correct. Only Technician A is correct. SRS connectors are usually yellow.

    Answer B is incorrect. The SRS module should be allowed to power down for 10 minutes or according to specs.

    Answer C is incorrect. Only Technician A is correct

    Answer D is incorrect. Only Technician B is incorrect.

21. A power steering pump is being replaced on a vehicle. Technician A says all the power steering hoses should also be replaced. Technician B says, depending on the system, the power steering filter should be cleaned or replaced. Who is correct?

    TASK D.5

    A. A only

    B. B only

    C. Both A and B

    D. Neither A nor B

    Answer A is incorrect. Total hose replacement may not be necessary.

    Answer B is correct. Only Technician B is correct. There may be a PS fluid filter in the system to be cleaned or replaced.

    Answer C is incorrect. Only Technician B is correct.

    Answer D is incorrect. Only Technician A is incorrect.

22. Technician A says that if a strut rod bushing has deteriorated, it is easier to replace the entire strut rod. Technician B says that strut rods themselves seldom fail. Who is correct?

A. A only

B. B only

C. Both A and B

D. Neither A nor B

Answer A is incorrect. Only the bushing need be replaced.

Answer B is correct. Only Technician B is correct. Strut rods seldom fail.

Answer C is incorrect. Only Technician B is correct.

Answer D is incorrect. Only Technician A is incorrect.

23. Which of the following parts, as shown in the illustration above, is being inspected for wear?

A. A shock absorber bushing

B. A strut bushing

C. A spring shackle

D. A spring bushing

Answer A is incorrect. This is not a shock bushing.

Answer B is incorrect. A strut bushing is not pictured.

Answer C is incorrect. This is not a spring shackle, but part of the spring itself.

Answer D is correct. The item shown is a leaf spring bushing.

24. When driving slowly over irregular surfaces, a clunking noise is heard coming from the front end of a vehicle. Technician A says that the vehicle's sway bar bushings may be faulty. Technician B says that the coil springs may be weak. Who is correct?

    TASK D.22

    A. A only
    B. B only
    C. Both A and B
    D. Neither A nor B

    Answer A is correct. Only Technician A is correct. Sway bar bushings exhibit this noise when worn.

    Answer B is incorrect. Coil springs would not cause this complaint.

    Answer C is incorrect. Only Technician A is correct

    Answer D is incorrect. Only Technician B is incorrect.

25. A vehicle is experiencing excessive rear-end sway on turns. Which of the following is the most likely cause?

    TASK D.29

    A. A broken sway bar link
    B. A faulty rear universal joint
    C. A faulty rear wheel bearing
    D. A broken axle flange

    Answer A is correct. A broken link would keep the sway bar from working properly.

    Answer B is incorrect. A rear U-joint would not affect sway.

    Answer C is incorrect. A faulty rear bearing would not cause vehicle sway.

    Answer D is incorrect. A broken axle flange would cause far more serious issues than sway.

26. Technician A says that a faulty rear ball joint could affect a vehicle's rear wheel alignment. Technician B says that a faulty rear ball joint could be a safety hazard. Who is correct?

    TASK D.34

    A. A only
    B. B only
    C. Both A and B
    D. Neither A nor B

    Answer A is incorrect. Both Technicians A and B are correct.

    Answer B is incorrect. Both Technicians A and B are correct.

    Answer C is correct. Both Technicians are correct. The rear wheel alignment would likely be affected, and it would be a definite safety hazard if it came apart!

    Answer D is incorrect. Both Technicians A and B are correct

27. A mist of oil is seen on the tubes of a vehicle's rear shock absorbers. The shocks:

    TASK D.37

    A. are normal.
    B. will need to be replaced with in a few thousand miles.
    C. should be replaced as soon as possible.
    D. should be replaced immediately.

    Answer A is correct. This is a normal condition.

    Answer B is incorrect. A mist of oil is normal.

    Answer C is incorrect. They do not need to be replaced.

    Answer D is incorrect. The shocks are fine as they are.

TASK D.39

28. With the steering wheel straight, a vehicle tends to drift to the left while travelling on a level road. Technician A says that the air pressure in the vehicle's right front tire may be low. Technician B says that the vehicle may need a wheel alignment. Who is correct?

A. A only

B. B only

C. Both A and B

D. Neither A nor B

Answer A is incorrect. The left front tire could low on air as the cause, but not the right front tire.

Answer B is correct. Only Technician B is correct. A wheel alignment may be needed.

Answer C is incorrect. Only Technician B is correct

Answer D is incorrect. Only Technician A is incorrect

TASK D.40

29. When a vehicle is being test driven straight on a level road, the steering wheel is not centered. Which of the following is the most likely cause?

A. The clock spring is faulty.

B. The steering wheel is incorrectly installed.

C. The steering wheel is bent.

D. A front strut is worn.

Answer A is incorrect. This would not cause the steering wheel to be off center.

Answer B is correct. The steering wheel may be installed one or more splines off center on the steering shaft.

Answer C is incorrect. This would be highly unlikely, plus it would not cause the problem described.

Answer D is incorrect. A worn strut would affect ride, but not cause the wheel to be off center.

TASK D.46

30. A vehicle experiences "dog tracking." Which of the following is the most likely cause?

A. Rear axle misalignment

B. Worn rear shocks

C. Worn rear springs

D. Worn rear tires

Answer A is correct. A misaligned rear axle would cause dog tracking (incorrect thrust line).

Answer B is incorrect. Worn shocks would affect the ride, but not the thrust line.

Answer C is incorrect. Worn springs would affect the ride, but not the thrust line.

Answer D is incorrect. Worn tires would not affect the thrust line.

31. When driven at very slow speeds, a vehicle's steering wheel tends to shimmy. Which of the following is the most likely cause?

    A. A faulty shock

    B. A faulty spring

    C. A faulty control arm bushing

    D. A faulty tire

TASK D.51

Answer A is incorrect. A faulty shock would not cause steering wheel shimmy.

Answer B is incorrect. A faulty spring would not cause steering wheel shimmy.

Answer C is incorrect. A faulty control arm bushing would not contribute to steering wheel shimmy.

Answer D is correct. A tire with a separated tread or other fault could be detected in the steering wheel.

32. Technician A says that it's OK to repair a tire when two puncture repairs overlap each other. Technician B says when two tire punctures are within several inches of each other and on the same tread line, the tire should be scrapped. Who is correct?

    A. A only

    B. B only

    C. Both A and B

    D. Neither A nor B

TASK D.56

Answer A is incorrect. Two repairs may not overlap.

Answer B is correct. Only Technician B is correct. Two repairs close to each other on the same tread line are not permitted; in such a case the tire should be scrapped.

Answer C is incorrect. Only Technician B is correct.

Answer D is incorrect. Only Technician A is incorrect.

33. During a test drive, a high-pitched metallic scraping noise is heard whenever a vehicle turns a corner. Which of these is the most likely cause?

    A. Faulty wheel bearings

    B. Faulty CV joints

    C. Worn brake pads

    D. Worn ball joints

TASK E.1

Answer A is incorrect. Faulty wheel bearings would likely produce a growling sound on turns.

Answer B is incorrect. Faulty CV joints would "click" on turns.

Answer C is correct. The brake pad wear indicators are doing their job.

Answer D is incorrect. Worn ball joints could "knock" on bumps, but not just on turns.

TASK E.4

34. A vehicle's red brake warning light stays on whenever the ignition is ON and the engine is running. Any of the following could be the cause EXCEPT:

A.   The parking brake is engaged.

B.   The brake fluid is low in the reservoir.

C.   The parking brake switch is out of adjustment.

D.   The ABS has a faulty wheel sensor.

Answer A is incorrect. A parking brake partially or fully ON would cause the red brake light to come on.

Answer B is incorrect. Low brake fluid would cause the red brake light to come on.

Answer C is incorrect. A maladjusted switch could cause the red brake light to come on.

Answer D is correct. A faulty wheel sensor would cause the yellow ABS brake light to come on.

TASK E.6

35. Technician A says it's a good idea to open a caliper bleed screw before retracting the caliper piston. Technician B says to start brake bleeding at the wheel closest to the master cylinder. Who is correct?

A.   A only

B.   B only

C.   Both A and B

D.   Neither A nor B

Answer A is correct. Only Technician A is correct. It's a good idea to avoid pushing contaminated fluid up into the brake system.

Answer B is incorrect. Brake bleeding generally starts the farthest away from the master cylinder.

Answer C is incorrect. Only Technician A is correct.

Answer D is incorrect. Only Technician B is incorrect.

TASK E.10

36. Technician A says that whenever drum brake shoes are being replaced, the drum brake hardware should be cleaned in a parts washer and re-used. Technician B says that when installing duo-servo brakes, the larger of the two shoes should be installed towards the front of the vehicle. Who is correct?

A.   A only

B.   B only

C.   Both A and B

D.   Neither A nor B

Answer A is incorrect. The parts should be replaced, not cleaned and reused.

Answer B is incorrect. The larger (secondary) shoe goes towards the rear of the car.

Answer C is incorrect. Neither Technician is correct

Answer D is correct. Neither Technician is correct.

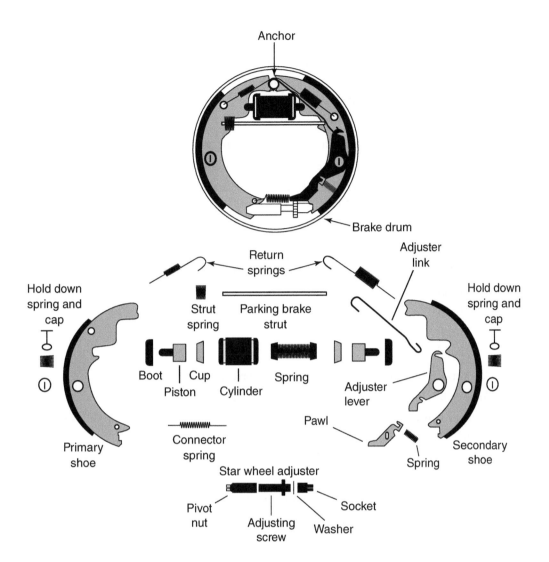

37. Which of the parts as shown in the illustration above, should be lubricated with water resistant grease during drum brake service?

    A. The parking brake strut and the adjuster lever

    B. The strut spring, connector spring and return springs

    C. The brake cylinder cups and boots

    D. The hold down spring and cap

TASK E.11

Answer A is correct. The contact points of these parts should be lubricated.

Answer B is incorrect. The springs do not need to be lubricated.

Answer C is incorrect. The cups are lubricated by the brake fluid itself; the boots are not to be lubricated.

Answer D is incorrect. These parts do not require lubrication.

38. A vehicle's parking brake should be adjusted:

A. before the service brakes are adjusted.

B. after the service brakes are adjusted.

C. with the parking brake lever applied.

D. with the ignition turned ON.

Answer A is incorrect. the parking brake should be adjusted after the service brakes.

Answer B is correct. Adjust the parking brake after the service brakes.

Answer C is incorrect. If made with the parking brake lever applied, the parking brake adjustment would be incorrect.

Answer D is incorrect. The ignition does not need to be ON to make the adjustment.

39. Technician A says that some disc brake pistons may be retracted using a "C" clamp. Technician B says that some caliper pistons may be retracted using a special retracting tool. Who is correct?

A. A only

B. B only

C. Both A and B

D. Neither A nor B

Answer A is incorrect. Technician B is also correct.

Answer B is incorrect. Technician A is also correct.

Answer C is correct. Both Technicians are correct. Either a "C" clamp or a special retracting tool may be used.

Answer D is incorrect. Neither Technician is incorrect.

40. Upon removal of a vehicle's left-front brake pads, one pad in the floating-type caliper is worn much farther than the other. Which of the following is the most likely cause?

A. A partially plugged brake hose

B. Rusted caliper pins or slides

C. A rusty caliper anchor plate

D. A sticking caliper piston

Answer A is incorrect. In this case, both pads would wear.

Answer B is correct. The caliper cannot correctly slide to remain centered.

Answer C is incorrect. Floating type calipers do not use an anchor plate.

Answer D is incorrect. A sticking piston would affect both brake pads.

41. Which of the following is used to check for rotor runout?

A. An inside micrometer

B. An outside micrometer

C. A dial indicator

D. A Vernier caliper

Answer A is incorrect. A micrometer would not show runout.

Answer B is incorrect. An outside micrometer would not show runout, only thickness.

Answer C is correct. A dial indicator could be useful for revealing wobble or runout of a rotor.

Answer D is incorrect. A Vernier caliper is not accurate enough, nor does it show runout.

42. When refilling a brake master cylinder, always use:

TASK E.27

    A. DOT 3 brake fluid.
    B. DOT 5 brake fluid.
    C. Synthetic brake fluid.
    D. OEM specified brake fluid.

Answer A is incorrect. Though used the most, DOT 3 may not always be the correct type.

Answer B is incorrect. DOT 5 may not be the correct type.

Answer C is incorrect. Synthetic fluid may not be the correct type.

Answer D is correct. Always follow the OEM recommendations when adding brake fluid.

43. A vehicle which uses the same hydraulic pressure pump for both the power brakes and power steering has a:

TASK E.34

    A. Bendix® brake system.
    B. Hydro-boost brake system.
    C. Vacu-boost brake system.
    D. Teves® brake system.

Answer A is incorrect. Bendix® describes a type of brakes, not the hydraulic system used.

Answer B is correct. The so-called "hydro-boost system" uses the same pump for both systems.

Answer C is incorrect. What is a vacu-boost system?

Answer D is incorrect. Teves® is a Tier 1 brake supplier; it is not a system.

44. Technician A says that current can be measured in a live circuit by connecting an ammeter in parallel across a load. Technician B says that current can be measured in a non-live (dead) circuit by using an amps clamp. Who is correct?

TASK F.3

    A. A only
    B. B only
    C. Both A and B
    D. Neither A nor B

Answer A is incorrect. An ammeter in parallel would short the circuit.

Answer B is incorrect. The circuit must be live to measure amperage flow.

Answer C is incorrect. Neither Technician is correct.

Answer D is correct. Neither Technician is correct.

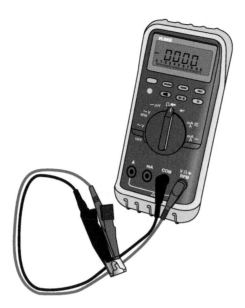

TASK F.4

45. As shown in the illustration above, a fuse is being checked for continuity with a DMM. In this case, the fuse:

A. is blown and should be replaced.

B. has high resistance and should be replaced.

C. has low resistance and should be cleaned.

D. has no resistance and is good as is.

Answer A is incorrect. The fuse is good.

Answer B is incorrect. The meter shows zero resistance.

Answer C is incorrect. The fuse has zero resistance.

Answer D is correct. The meter reads zero; i.e. no resistance.

TASK F.5

46. A battery load test is being performed. The proper method is to:

A. apply a load equal to twice the CCA rating.

B. apply a load equal to one-half the CCA rating.

C. apply a load for 30 seconds.

D. apply a load until the battery drops to 9.6 volts.

Answer A is incorrect. This is an incorrect procedure.

Answer B is correct. This is the correct procedure.

Answer C is incorrect. The load should be applied for 15 seconds.

Answer D is incorrect. The load should be applied for 15 seconds, after which take note of the voltage.

47. A 12-volt lead acid battery is being charged. Technician A says that when connecting the battery charger to a battery the cell caps should remain on the battery. Technician B says that when connecting the battery charger to the battery safety glasses should be worn. Who is correct?

TASK F.12

  A. A only
  B. B only
  C. Both A and B
  D. Neither A nor B

  Answer A is correct. The battery caps should remain on the battery during charging because they contain flame arrestors.

  Answer B is incorrect. Safety goggles and a face shield should be worn for adequate eye protection.

  Answer C is incorrect. Technician A is correct.

  Answer D is incorrect. Technician A is correct.

48. A load test is being performed on a vehicle's charging system. Technician A says it's OK to increase the load until the charging system voltage just starts to drop, then take note the amperage. Technician B says it's OK to apply a load until the alternator light comes on, then take note of the amperage. Who is correct?

TASK F.14

  A. A only
  B. B only
  C. Both A and B
  D. Neither A nor B

  Answer A is incorrect. Neither Technician is correct.

  Answer B is incorrect. This is not the correct procedure; it would likely damage the alternator.

  Answer C is incorrect. Neither Technician is correct.

  Answer D is correct. Neither Technician is correct. Run engine at manufacturers recommended rpm. Adjust carbon pile on load tester to obtain maximum current output while maintaining voltage above 12 volts. Amps should be within 10% of rated output.

TASK F.15

49. The puller shown above is used to remove a(an):

    A. ball joint.

    B. tie rod end.

    C. alternator decoupling pulley.

    D. steering knuckle.

    Answer A is incorrect. This tool does not remove ball joints.

    Answer B is incorrect. This tool does not remove tie rod ends.

    Answer C is correct. The tool shown is used to remove an alternator decoupling pulley (ADP).

    Answer D is incorrect. This tool is not used to remove a steering knuckle.

TASK F.17

50. A headlight alignment procedure is being performed. Which of the following is LEAST LIKELY to cause the alignment to be incorrect?

    A. A sloping drive surface

    B. Low tire pressure

    C. dirty headlight lenses

    D. A heavy load in the vehicle trunk

    Answer A is incorrect. The headlight alignment would likely be faulty.

    Answer B is incorrect. The tires should be correctly inflated.

    Answer C is correct. This is an EXCEPT type question. Dirty headlights would not contribute to a headlight misalignment.

    Answer D is incorrect. The vehicle should be loaded with a typical load, not an excess.

51. Technician A says that the brake warning light on the instrument panel cluster dash is a yellow light. Technician B says that the ABS indicator light on the IPC is red. Who is correct?

   A. A only
   B. B only
   C. Both A and B
   D. Neither A nor B

TASK F.20

   Answer A is incorrect. The brake warning light is red.

   Answer B is incorrect. The ABS light is yellow.

   Answer C is incorrect. Neither Technician is correct.

   Answer D is correct. Neither Technician is correct.

52. With the vehicle engine running, the A/C compressor fails to engage when the A/C switch is turned ON. Which of the following is the most likely cause?

   A. A faulty blower motor
   B. A faulty A/C relay
   C. A faulty blend door sensor
   D. A faulty heater core

TASK G.1

   Answer A is incorrect. A faulty blower would not likely affect the A/C compressor.

   Answer B is correct. A faulty relay would prevent A/C clutch engagement.

   Answer C is incorrect. A faulty blend door function would not affect the compressor.

   Answer D is incorrect. A faulty heater core would not affect the compressor.

53. During an underhood inspection for an inoperative A/C system, a technician notices an oily residue on the fins of the cooling system radiator. Which of the following could be the cause?

   A. A faulty compressor
   B. A faulty condenser
   C. A faulty evaporator
   D. A faulty expansion valve

TASK G.2

   Answer A is incorrect. A compressor would not deposit oil on the radiator.

   Answer B is correct. A leaking condenser could leak (compressor) oil which might show on the radiator behind it.

   Answer C is incorrect. A faulty evaporator compressor would not deposit oil on the radiator.

   Answer D is incorrect. A faulty expansion valve would not deposit oil on the radiator.

54. Technician A says that leaves caught in the condenser could cause the cabin heating system to operate poorly. Technician B says that leaves caught in the condenser could cause the A/C system to operate poorly. Who is correct?

   A. A only
   B. B only
   C. Both A and B
   D. Neither A nor B

TASK G.3

   Answer A is incorrect. A blocked condenser would not keep the heating system from delivering warm air.

   Answer B is correct. Only Technician B is correct. A blocked condenser would not cool the refrigerant properly.

   Answer C is incorrect. Only Technician B is correct.

   Answer D is incorrect. Only Technician A is incorrect.

TASK G.5

55. Technician A says that the best way to determine if a newer style serpentine belt is worn is to visually inspect it. Technician B says that the best way to determine if a newer style serpentine belt is worn is to check it with a wear gauge. Who is correct?

    A. A only

    B. B only

    C. Both A and B

    D. Neither A nor B

Answer A is incorrect. Newer style serpentine belts do not necessarily show wear as one would expect.

Answer B is correct. Only Technician B is correct. A belt wear gauge will show belt groove wear.

Answer C is incorrect. Only Technician B is correct.

Answer D is incorrect. Only Technician A is incorrect.

PREPARATION EXAM 2—ANSWER KEY

1.	C	**20.**	D	**39.**	B
2.	D	**21.**	B	**40.**	B
3.	D	**22.**	B	**41.**	B
4.	C	**23.**	A	**42.**	C
5.	C	**24.**	D	**43.**	D
6.	D	**25.**	C	**44.**	C
7.	C	**26.**	A	**45.**	A
8.	A	**27.**	C	**46.**	B
9.	C	**28.**	C	**47.**	D
10.	A	**29.**	C	**48.**	D
11.	B	**30.**	D	**49.**	D
12.	D	**31.**	C	**50.**	C
13.	D	**32.**	D	**51.**	C
14.	A	**33.**	D	**52.**	D
15.	A	**34.**	A	**53.**	B
16.	C	**35.**	D	**54.**	C
17.	D	**36.**	D	**55.**	D
18.	A	**37.**	A		
19.	B	**38.**	B		

PREPARATION EXAM 2—EXPLANATIONS

1. A high mileage engine has a leaking oil pan gasket. Technician A says some undercar components may need to be removed in order to replace the pan gasket. Technician B says that the engine may have to be raised up off its engine mounts to replace the gasket. Who is correct?

TASK A.2

   A. A only

   B. B only

   C. Both A and B

   D. Neither A nor B

   Answer A is incorrect. Technician B is also correct.

   Answer B is incorrect. Technician A is also correct.

   Answer C is correct. Both Technicians are correct. Sometimes other parts may have to be removed in order to R&R the oil pan and replace a leaking gasket. This may also mean raising the engine off its mounts.

   Answer D is incorrect. Both Technicians are making correct statements.

2. Oil pan and valve cover gasket(s) should be checked for:

TASK A.4

   A. rust.

   B. stains.

   C. corrosion.

   D. leaks.

   Answer A is incorrect. Gaskets do not rust.

   Answer B is incorrect. Stains do not affect gaskets.

   Answer C is incorrect. Gaskets do not suffer from corrosion.

   Answer D is correct. Gaskets do leak (more often than desired).

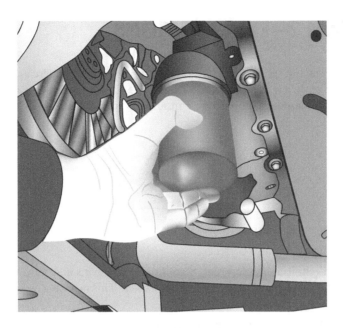

TASK A.5

3. The technician shown in the illustration above is changing the:

A. gasoline filter.

B. transmission filter.

C. differential filter.

D. oil filter.

Answer A is incorrect. A gasoline filter is not shown here.

Answer B is incorrect. The trans filter is usually inside or near the transmission.

Answer C is incorrect. The differential does not have a filter.

Answer D is correct. An oil filter is being changed (oil running down the fingers).

TASK A.7

4. Which of the following is the preferred method of checking a newer style serpentine belt for wear?

A. Twist the belt one-quarter turn and check for tightness.

B. Use a belt tension gauge.

C. Use a serpentine belt wear gauge.

D. Check the wear indicator on the pulley.

Answer A is incorrect. This does not check a belt for wear.

Answer B is incorrect. Belt tension is not necessarily an indicator of belt wear.

Answer C is correct. Because newer belt materials no longer show much wear and tear, manufacturers now recommend using a belt gauge to help determine if replacement is necessary.

Answer D is incorrect. Belt wear may not be associated with stretching or a loss of tension.

5. A vehicle's cooling system has been drained and flushed, and fresh coolant is being added to a vehicle. Technician A says to mix the correct ratio of coolant to water for the correct freeze point. Technician B says a bleed screw may have to be opened to purge air from the cooling system. Who is correct?

TASK A.10

    A. A only
    B. B only
    C. Both A and B
    D. Neither A nor B

    Answer A is incorrect. Technician B is also correct.

    Answer B is incorrect. Technician A is also correct.

    Answer C is correct. Both Technicians are correct.

    Answer D is incorrect. Neither Technician is incorrect.

6. Any routine inspection and maintenance program includes the periodic replacement of the:

TASK A.15

    A. transmission filter.
    B. power steering fluid filter.
    C. cabin filter.
    D. air filter.

    Answer A is incorrect. The transmission filter is seldom replaced; perhaps at 100,000 miles for normal vehicle use.

    Answer B is incorrect. The power steering (PS) filter is not often cleaned or replaced.

    Answer C is incorrect. The cabin filter is not often change, but it should be inspected every 15,000 miles according to many vehicle manufacturers.

    Answer D is correct. The intake air filter should be inspected and/or replaced more often than the other filters mentioned.

7. Which of the following should be used when replacing defective vehicle exhaust system components?

TASK A.17

    A. Galvanized exhaust system components
    B. Aluminum exhaust system components
    C. Stainless steel exhaust system components
    D. Titanium exhaust system components

    Answer A is incorrect. Today's exhaust systems are not made of galvanized metal.

    Answer B is incorrect. Exhaust systems are not made of aluminum.

    Answer C is correct. Today's vehicle exhaust systems are made of stainless steel to last the lifetime of the vehicle.

    Answer D is incorrect. Except for perhaps an exotic car, exhaust systems are not made of titanium due to cost or lack of need.

TASK A.18

8. When checking for diagnostic trouble codes (DTCs), it's important to:

    A. use the proper scan tool and software for the vehicle under test.
    B. clear all the codes before performing any tests.
    C. check the vehicle owner's manual before performing any tests.
    D. take the vehicle for a test drive to run all the monitors.

    Answer A is correct. Always use the proper scan tools and specifications when checking vehicle fault codes (DTCs).

    Answer B is incorrect. Fault codes should not be cleared until recorded and verified.

    Answer C is incorrect. The owner's manual would not contain the needed service or testing information.

    Answer D is incorrect. The monitors would have been run and the fault codes previously recorded. The service technician would now check Freeze Frame data in the PCM instead.

TASK A.20

9. On an OBD II vehicle, an EVAP leak may be associated with any of the following events EXCEPT:

    A. the MIL will be illuminated.
    B. a rubber hose may be split.
    C. carbon monoxide will be released to the atmosphere.
    D. the gas cap will allow VOCs to leak to the atmosphere.

    Answer A is incorrect. This is a true statement; the MIL would be ON.

    Answer B is incorrect. This is also a true statement; a slip hose could cause a minor or a major leak and trip a code.

    Answer C is correct. CO is not released from the EVAP system; however, VOCs could be released.

    Answer D is incorrect. This is quite commonly the cause of an EVAP-related MIL, so check the gas cap first.

TASK B.1

10. Which of the following is the LEAST LIKELY way a technician would confirm that an automatic transmission is faulty?

    A. Run the engine with the vehicle in NEUTRAL.
    B. Interview the customer.
    C. Take the vehicle for a test drive.
    D. Check for fault codes in the TCM.

    Answer A is correct. This is not a reliable way to uncover automatic transmission faults.

    Answer B is incorrect. True statement. In some cases a customer interview might prove helpful when determining what the complaint is.

    Answer C is incorrect. A test drive would be very helpful to verify a complaint.

    Answer D is incorrect. True statement; pull codes to find out what is hurting the vehicle.

11. Technician A says that the technician shown in the illustration above, could be checking the differential fluid level. Technician B says that the technician shown in this illustration above, could be checking the transmission fluid level. Who is correct?

TASK B.2

A. A only

B. B only

C. Both A and B

D. Neither A nor B

Answer A is incorrect. A differential does not have a dipstick.

Answer B is correct. Only Technician B is correct. The technician is in fact checking the transmission fluid level on the dipstick.

Answer C is incorrect. Only Technician B is making a correct statement.

Answer D is incorrect. Only Technician A is making an incorrect statement.

12. A technician is inspecting a vehicle's rear transmission mount and finds its rubber is dry and has split. The mount should be:

TASK B.6

A. replaced it with a solid steel mount.

B. replaced with a synthetic mount.

C. replaced with similar used part.

D. replaced with a new OEM or aftermarket part.

Answer A is incorrect. A solid mount would transfer noise to the vehicle chassis.

Answer B is incorrect. If even available, it may not be the recommended part for the vehicle.

Answer C is incorrect. A new replacement should be used.

Answer D is correct. The replacement part should be a new and correct for the vehicle.

13. Technician A says that all automatic transmission fluids (ATF) are interchangeable. Technician B says that new ATF should be pink in color. Who is correct?

    A. A only
    B. B only
    C. Both A and B
    D. Neither A nor B

    Answer A is incorrect. There are many types of different ATF.

    Answer B is incorrect. ATF is normally red in color.

    Answer C is incorrect. Neither Technician is making a correct statement.

    Answer D is correct. Neither Technician is correct.

14. Following drivetrain repairs to a vehicle, a final test drive reveals a vibration problem which increases with road speed and engine torque. Which of the following is the most likely cause?

    A. Incorrect drive line phasing
    B. Over-torqued wheel lug nuts
    C. Excessive pinion gear lash
    D. Loose spring shackles

    Answer A is correct: An out-of-phase driveshaft would cause a vibration complaint.

    Answer B is incorrect. This would not cause the stated complaint.

    Answer C is incorrect. Gear lash would not cause the stated complaint.

    Answer D is incorrect. Loose shackles would not cause the stated complaint.

15. While a brake shoe replacement procedure is being performed, gear lube is seen leaking from behind the rear axle flange. Technician A says that a leaking axle seal could be the cause. Technician B says that a faulty brake drum seal could be the cause. Who is correct?

    A. A only
    B. B only
    C. Both A and B
    D. Neither A nor B

    Answer A is correct. Only Technician A is correct. A leaking axle seal is the likely cause.

    Answer B is incorrect. A brake drum seal retains bearing grease, not gear lube.

    Answer C is incorrect. Only Technician A is correct.

    Answer D is incorrect. Only Technician B is incorrect.

16. A wheel stud needs to be replaced on a solid rear axle flange. Technician A says that on some vehicles the axle may have to be removed to make installing the replacement wheel stud possible. Technician B says that on some vehicles it may be possible to install a new wheel stud without removing the axle. Who is correct?

TASK C.12

   A. A only
   B. B only
   C. Both A and B
   D. Neither A nor B

Answer A is incorrect. Technician B is also correct

Answer B is incorrect. Technician A is also correct

Answer C is correct. Both Technicians are correct. On larger vehicles, it may not be possible to insert the new stud into the flange from behind without first sliding out the axle. On some vehicles it may be possible to insert the wheel stud from behind the flange.

Answer D is incorrect. Both Technicians are correct.

17. A FWD vehicle experiences a clicking noise from the left front wheel area during slow left-hand turns. Which of the following is the most likely cause?

TASK C.16

   A. A noisy brake pad
   B. Worn ball joints
   C. Worn sway bar bushings
   D. A faulty CV joint

Answer A is incorrect. A brake pad would not make a "click" noise.

Answer B is incorrect. A ball joint would not click.

Answer C is incorrect. A loose sway bar would not click.

Answer D is correct. This is a common sign that a CV joint is faulty and needs to be replaced.

18. A sealed front wheel bearing is to be installed. Technician A says the hub may have to be removed in order to install the bearing. Technician B says the bearing may have to be packed before installation. Who is correct?

TASK C.18

   A. A only
   B. B only
   C. Both A and B
   D. Neither A nor B

Answer A is correct. Only Technician A is correct. The hub will likely need to be removed in order to press in the bearing.

Answer B is incorrect. It's a sealed bearing.

Answer C is incorrect. Only Technician A is correct.

Answer D is incorrect. Only Technician B is incorrect.

TASK C.19

19. Which of the following is true about the fluid used in a transfer case?

A.   It is the same as the gear lube used in the differential.

B.   It may be unique because of specially designed additives.

C.   It is the same as automatic transmission fluid (ATF).

D.   It is the same as that used in manual transmissions.

Answer A is incorrect. Transfer case fluid may have different additives.

Answer B is correct. Special blends and additives may be used in transfer case fluid.

Answer C is incorrect. Transfer case fluid is not the same as ATF.

Answer D is incorrect. The fluid is sometimes specially formulated for the transfer case.

TASK D.2

20. Any of the following would likely indicate the correct power steering fluid to be used in a vehicle EXCEPT:

A.   the owner's manual.

B.   the service manual.

C.   a parts and labor guide.

D.   a service bulletin.

Answer A is incorrect. The owner's manual would specify the correct PS fluid in the maintenance section.

Answer B is incorrect. The service manual is a reliable source for this information.

Answer C is incorrect. The Parts section would provide the correct part number for the PS fluid.

Answer D is correct. Service bulletins are not typically used to convey routine maintenance information.

TASK D.4

21. A vehicle exhibits a groaning noise whenever the steering wheel is turned. Which of the following is the most likely cause?

A.   The system is overfilled with fluid.

B.   The system needs to be purged of air.

C.   The power steering pump is faulty.

D.   The power steering rack is faulty.

Answer A is incorrect. This would not cause the problem described.

Answer B is correct. A groaning noise is symptomatic of air in the PS system.

Answer C is incorrect. The pump is likely OK.

Answer D is incorrect. The rack is also likely OK.

22. During an undercar inspection, a bent lower control arm is discovered. Technician A says it may be heated and straightened. Technician B says that after the appropriate repair, the wheels should be realigned. Who is correct?

TASK D.13

A. A only

B. B only

C. Both A and B

D. Neither A nor B

Answer A is incorrect. The part should be replaced.

Answer B is correct. Only Technician B is correct. Changing suspension components will likely alter the vehicle's wheel alignment.

Answer C is incorrect. Only Technician B is correct.

Answer D is incorrect. Only Technician A is incorrect.

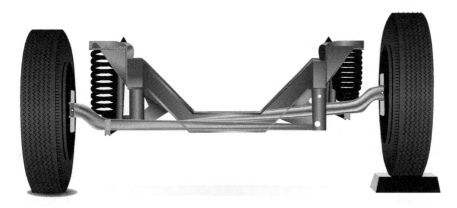

23. An older model pickup with a twin I-beam front suspension as shown above is found to need a camber adjustment. Technician A says that the I-beam(s) may need to be bent to achieve the alignment. Technician B says the radius rods may need to be bent to achieve the alignment. Who is correct?

TASK D.17

A. A only

B. B only

C. Both A and B

D. Neither A nor B

Answer A is correct. Only Technician A is correct. Though seemingly drastic, this is how the alignment may be performed.

Answer B is incorrect. The radius rods would not affect camber.

Answer C is incorrect. Only Technician A is correct.

Answer D is incorrect. Only Technician B is incorrect.

TASK D.19

24. The front-end coil springs on a vehicle have sagged. Which of the following would be most affected?

    A. Caster

    B. Toe

    C. Thrust line

    D. Ride height

    Answer A is incorrect. Caster would not likely be affected by weak springs.

    Answer B is incorrect. Toe would not be affected.

    Answer C is incorrect. The thrust line would not be affected.

    Answer D is correct. Weak springs would directly affect ride height.

TASK D.27

25. The rear coil spring on a vehicle is found to be broken. Which of the following is the LEAST LIKELY cause?

    A. The vehicle was overloaded.

    B. The vehicle was used for off-roading.

    C. The vehicle was not well maintained.

    D. The vehicle was used for heavy towing.

    Answer A is incorrect. Overloading a vehicle could result in breaking a spring.

    Answer B is incorrect. Off road driving of a vehicle could break a spring.

    Answer C is correct. There is no routine maintenance to be performed on rear springs.

    Answer D is incorrect. Towing with excessive tongue weight could cause spring damage.

TASK D.31

26. The rear rebound bumpers of a vehicle are found to be cracked and collapsed. Technician A says that routine vehicle overloading could be the cause. Technician B says that once the bumpers are replaced, a wheel alignment will be needed. Who is correct?

    A. A only

    B. B only

    C. Both A and B

    D. Neither A nor B

    Answer A is correct. Only Technician A is correct. Serious overloading combined with rough road use could cause the bumpers to be ruined.

    Answer B is incorrect. Replacing the bumpers does not affect alignment.

    Answer C is incorrect. Only Technician A is correct.

    Answer D is incorrect. Only Technician B is incorrect.

27. Upon inspection, the rear tie rods of an independently suspended vehicle are found to have 1/4 inch of play. Technician A says that they should be replaced. Technician B says that if tie rods are replaced, a wheel alignment will be required. Who is correct?

TASK D.35

   A.   A only

   B.   B only

   C.   Both A and B

   D.   Neither A nor B

   Answer A is incorrect. Generally speaking, any play beyond 1/8 inch requires replacement of the tie rod end.

   Answer B is incorrect. Replacing tie rod ends will upset the wheel alignment.

   Answer C is correct. Both Technicians are correct.

   Answer D is incorrect. Neither Technician is incorrect.

28. A vehicle tends to pull to one side of the road when travelling on a straight road. Which of the following is the most likely cause?

TASK D.39

   A.   Incorrect steering wheel indexing/installation

   B.   Incorrect thrust line

   C.   Incorrect camber setting

   D.   A faulty steering damper

   Answer A is incorrect. Steering wheel indexing on the steering shaft would not affect a vehicle's directional ability.

   Answer B is incorrect. An incorrect thrust line would not significantly affect directional ability.

   Answer C is correct. The camber setting directly affects straight and true vehicle travel.

   Answer D is incorrect. The condition of the steering damper does not affect directional ability.

29. Technician A says that caster on a short-long arm (SLA) equipped vehicle may be adjusted by adding or subtracting shims. Technician B says that caster on a SLA equipped vehicle may be adjusted by rotating an eccentric shaft. Who is correct?

TASK D.42

   A.   A only

   B.   B only

   C.   Both A and B

   D.   Neither A nor B

   Answer A is incorrect. Technician B is also correct.

   Answer B is incorrect. Technician A is also correct.

   Answer C is correct. Both Technicians are correct.

   Answer D is incorrect. Neither Technician is incorrect.

30. A tire is found to have a nail hole in its sidewall. Which of the following should be done?

   A. Repair the tire with a string type plug.

   B. Repair the tire with a patch from the inside.

   C. Repair the tire with a vulcanizing kit.

   D. Scrap and replace the tire with a similar good one.

   Answer A is incorrect. A sidewall puncture is not to be repaired.

   Answer B is incorrect. A sidewall puncture is not repairable.

   Answer C is incorrect. A sidewall puncture means scrapping the tire.

   Answer D is correct. The tire should be scrapped because a safe repair cannot be performed to the sidewall.

31. The four non-asymmetric/non-directional radial tires on a 4WD vehicle with a mini-spare tire are to be rotated. Which of the following is the correct procedure?

   A. Exchange both front and rear tires from side to side.

   B. Exchange the front and rear tires with each other on the same side.

   C. Move the rear tires to the front and the front tires to the opposite rear.

   D. Move the rear tires to the opposite front and the front tires to the rear.

   Answer A is incorrect. This is the correct procedure for a vehicle with two radial and two conventional tires.

   Answer B is incorrect. This is the correct procedure for a vehicle with directional tires.

   Answer C is correct. This is the correct procedure to be followed on a 4WD vehicle with a mini-spare tire.

   Answer D is incorrect. This is the correct procedure for a front-wheel drive (FWD) vehicle.

32. A vehicle with the same make/size of tires and wheels and the same recommended inflation on all 4 wheels experiences an illuminated TPMS light. Technician A says that the vehicle's tires may have been rotated. Technician B says that a TPMS relearn procedure was not performed. Who is correct?

   A. A only

   B. B only

   C. Both A and B

   D. Neither A nor B

   Answer A is incorrect. Tire rotation on this TPMS equipped vehicle will not cause the TPMS light to come on.

   Answer B is incorrect. A relearn procedure should be performed after a tire rotation is performed, but failure to do so in this case would not cause the TPMS light to illuminate.

   Answer C is incorrect. Neither Technician is correct. While tire rotation on a TPMS equipped vehicle does require a relearn procedure, simply rotating the tires on this vehicle will not cause the TPMS light to illuminate because all tire/wheel assemblies are of the same size and pressure.

   Answer D is correct. Neither Technician is correct.

33. The yellow ABS warning light is illuminated on a vehicle's instrument panel. Which of the following is the most likely cause?

    A. A wheel is out of balance.

    B. A tire is defective.

    C. A TPMS sending unit is faulty.

    D. A tone wheel was damaged during a tire repair.

TASK E.4

Answer A is incorrect. A wheel imbalance should not affect the ABS

Answer B is incorrect. A tire defect should not affect the ABS.

Answer C is incorrect. A TPMS sending unit should not affect ABS.

Answer D is correct. A faulty tone wheel will trip an ABS code and illuminate the ABS warning light.

34. A short to ground in the parking light circuit will likely cause:

    A. the brake warning light to stay on.

    B. the brake warning light to not illuminate.

    C. an MIL to be illuminated.

    D. a code to be set in the EBCM.

TASK E.5

Answer A is correct. A grounded circuit will activate the light.

Answer B is incorrect. The light would stay ON.

Answer C is incorrect. An MIL will not be caused to illuminate.

Answer D is incorrect. A code will not be set by a grounded park brake circuit.

35. The lid on a can of DOT-3 brake fluid has been left off its container overnight. Technician A says it's OK to replace the cap and continue to use the fluid. Technician B says the brake fluid may be filtered through cheese cloth and then used. Who is correct?

    A. A only

    B. B only

    C. Both A and B

    D. Neither A nor B

TASK E.7

Answer A is incorrect. Technician B is also incorrect.

Answer B is incorrect. Technician A is also incorrect.

Answer C is incorrect. Both Technicians are incorrect; the brake fluid should be discarded.

Answer D is correct. Neither Technician is correct. The brake fluid may have absorbed water or humidity; it should be safely discarded.

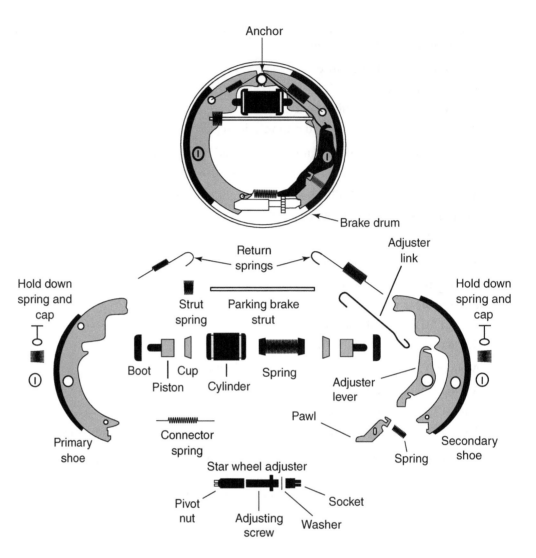

TASK E.13

36. Brake shoes are being installed at the rear of a vehicle. Which of the parts in the illustration above, should be lubricated with water resistant grease?

A. The brake cylinder cups

B. The hold downs

C. The shoe linings

D. The star wheel adjuster

Answer A is incorrect. Wheel cylinder cups are lubricated by the brake fluid.

Answer B is incorrect. The hold downs do not require lubrication.

Answer C is incorrect. The shoes' lining material should never have any lubricant on them.

Answer D is correct. The star wheel adjuster should be greased before installation.

37. A technician is attempting to retract a brake caliper piston by using a "C" clamp, but the piston will not retract. Which of the following is the most likely cause?

TASK E.17

   A. The caliper piston also serves as a parking brake.

   B. The caliper piston "O" ring seal is twisted.

   C. The caliper is a full floating type component.

   D. the caliper is a semi-floating type.

Answer A is correct. The piston must be rotated to retract it.

Answer B is incorrect. An "O" ring his would not cause the piston to bind.

Answer C is incorrect. Because it is a floating caliper does not mean the piston should be stuck.

Answer D is incorrect. This is also irrelevant to a stuck piston.

38. While inspecting the disc brakes of a vehicle, the technician finds that only the inner pad has worn down to its backing plate. Technician A says that the caliper piston should be inspected for possible damage from the pad. Technician B says that the rotor should be inspected for damage from the pad. Who is correct?

TASK E.20

   A. A only

   B. B only

   C. Both A and B

   D. Neither A nor B

Answer A is incorrect. Pad wear will not directly affect the caliper piston.

Answer B is correct. Only Technician B is correct. The rotor may have been scored by the worn down pad.

Answer C is incorrect. Only Technician B is correct

Answer D is incorrect. Only Technician A is incorrect.

39. During a routine disc brake inspection, the caliper slide pin boots are found to be torn. Which of the following should be done?

TASK E.21

   A. Replace the caliper pins and boots.

   B. Inspect the caliper pins for corrosion.

   C. Inspect the caliper pistons.

   D. Replace the caliper slides.

Answer A is incorrect. The pins may be OK.

Answer B is correct. The pins may have become rusted or corroded.

Answer C is incorrect. There is no relationship between the pins and the pistons

Answer D is incorrect. Slides are not used on this type of calipers.

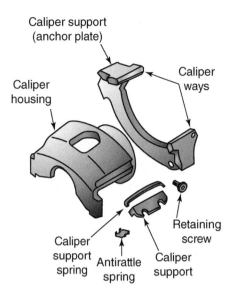

Caliper support (anchor plate)

Caliper housing

Caliper ways

Retaining screw

Caliper support spring

Antirattle spring

Caliper support

TASK E.25

40. The disc brake components illustrated above are being assembled. Which of the following part(s) should be lubricated with waterproof grease?

A. The caliper housing

B. The caliper ways

C. The support spring

D. The caliper anchor plate

Answer A is incorrect. The housing does not need lubrication.

Answer B is correct. The caliper anchor plate must slide at this area as the brake pads wear.

Answer C is incorrect. The support spring does not need lubrication.

Answer D is incorrect. The anchor plate does not need lubrication.

TASK E.32

41. Which of the following is the best method to use when checking a vacuum brake booster check valve for leaks?

A. Remove the valve and blow compressed air through it to see if it leaks.

B. Run the engine, then shut it off and pull the brake booster check valve to see if air rushes in.

C. Remove the valve and see if light passes through it.

D. Remove the vacuum line and attempt to blow air through it.

Answer A is incorrect. The valve could be damaged.

Answer B is correct. Vacuum should not escape from the booster until the check valve is removed.

Answer C is incorrect. Looking for light through the one-way check valve is not a recommended test.

Answer D is incorrect. The valve could be damaged.

42. An electric-hydraulic assist brake system is to be checked for leaks. Technician A says that the hydraulic hoses should be inspected for signs of leakage. Technician B says that the hose fittings should be inspected for signs of leakage. Who is correct?

TASK E.33

A. A only

B. B only

C. Both A and B

D. Neither A nor B

Answer A is incorrect. Technician B is also correct.

Answer B is incorrect. Technician A is also correct.

Answer C is correct. Both Technicians are correct.

Answer D is incorrect. Neither Technician is incorrect.

43. A hydraulic type brake booster system is to be inspected for faults. The inspection should start with all of the following EXCEPT:

TASK E.34

A. the brake fluid level.

B. the power steering belt.

C. the hose connections.

D. the master cylinder pushrod.

Answer A is incorrect. The fluid level should be checked.

Answer B is incorrect. The power steering drive belt should be checked.

Answer C is incorrect. Check all hose connections for leaks.

Answer D is correct. The pushrod is unlikely to be problematic; plus, inspecting it is labor intensive.

44. An air bag system (supplemental restraint system - SRS) needs to be disarmed before SRS related service is performed. Technician A says to pull the SRS fuse. Technician B says to disconnect the battery. Who is correct?

TASK F.1

A. A only

B. B only

C. Both A and B

D. Neither A nor B

Answer A is incorrect. Technician B is also correct.

Answer B is incorrect. Technician A is also correct.

Answer C is correct. Both Technicians are correct. Both of these procedures may be performed to disarm the SRS system.

Answer D is incorrect. Neither Technician is incorrect.

45. When measuring for load resistance in a circuit, the technician should be sure to:

 A. turn the circuit power off.

 B. have the circuit power on.

 C. connect the ohmmeter in series with the load.

 D. connect the ammeter in parallel with the load.

Answer A is correct. When using an ohmmeter, the circuit should be OFF; it should not be live.

Answer B is incorrect. An ohmmeter should seldom if ever be used in a live circuit.

Answer C is incorrect. An ohmmeter would not be connected in series.

Answer D is incorrect. An ammeter is not to be connected in parallel.

46. Before disconnecting a vehicle's 12-volt battery for service, providing auxiliary 12-volt power to the vehicle will accomplish all of these EXCEPT:

 A. make certain engine operating parameters are maintained in the PCM.

 B. pose the risk of damage to the PCM.

 C. maintain automatic power seat-position memory settings.

 D. keep radio station presets stored in keep alive memory (KAM).

Answer A is incorrect. Connecting auxiliary power to the vehicle before disconnecting the service battery will help retain engine operating parameters (such as short-term and long-term fuel trim settings) stored in the powertrain control module (PCM).

Answer B is correct. While a loss of battery power while re-flashing the PCM could cause permanent damage to it, simply disconnecting and removing the 12-volt battery does not pose such a risk.

Answer C is incorrect. Auxiliary power must be provided to the vehicle if power seat-position memory settings are to be retained.

Answer D is incorrect. Auxiliary power will assure that the vehicle's radio station presets are retained in keep alive memory (KAM).

47. Which of the following is the best procedure to follow when using a battery charger on a discharged battery?

 A. Use a fast charge with high amperage on flooded lead-acid batteries.

 B. Use a slow charge with high amperage on flooded lead-acid batteries.

 C. Use a slow charge with high voltage on valve regulated lead-acid batteries.

 D. Use the recommended volts/amps charge rate specified for the battery type.

Answer A is incorrect. Fast high-amp charging could overheat and ruin the battery.

Answer B is incorrect. Sustained high amperage could ruin the battery.

Answer C is incorrect. This may not be the correct method for VRLA batteries.

Answer D is correct. The manufacturer's recommended charge rate should be used.

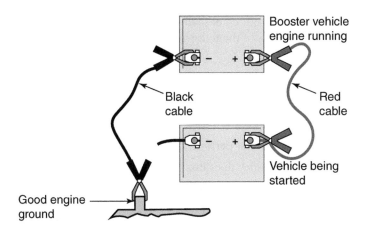

48. When attempting to jump-start a vehicle above, which of the following jumper cable connections should be made last?

    A. Red cable to the booster vehicle
    B. Red cable to the vehicle being started
    C. Black cable to the booster vehicle
    D. Black cable to a good engine ground

TASK F.10

Answer A is incorrect. This method presents a risk of a battery explosion.

Answer B is incorrect. This also presents a risk of a battery explosion.

Answer C is incorrect. This method is likewise unsafe and risks battery explosion.

Answer D is correct. This would be the safest means of jump starting. The last (and possible arcing) connection is away from the battery and would most likely prevent a battery explosion.

49. A starter motor is being removed from a vehicle. Technician A says it's important to disconnect the heavy battery wire from the starter solenoid and tape it to prevent a short circuit to a vehicle ground. Technician B says it's important to remove and tape the small wire lead to the starter solenoid to prevent it from shorting to a vehicle ground. Who is correct?

    A. A only
    B. B only
    C. Both A and B
    D. Neither A nor B

TASK F.13

Answer A is incorrect. The battery should be disconnected to prevent a short circuit to ground.

Answer B is incorrect. Likewise, the battery should always be disconnected to prevent a short circuit to ground.

Answer C is incorrect. Neither Technician is correct.

Answer D is correct. Neither Technician is correct. The battery should be disconnected instead.

50. An alternator pulley designed to smooth out crankshaft torque reversals otherwise imposed on the alternator is to be replaced. Which of the following should be used?

 A. Solid pulley

 B. Overrunning pulley

 C. Decoupling pulley

 D. Overdrive pulley

 Answer A is incorrect. A solid pulley does not perform this function.

 Answer B is incorrect. An overrunning pulley does not fully smooth out torque reversals.

 Answer C is correct. The Alternator Decoupling Pulley (ADP) performs this function.

 Answer D is incorrect. This is a fictitious part.

51. A heater control panel with an LED display fails to illuminate and the controls do not operate properly. Technician A says to check the fuse and all electrical connections. Technician B says the circuit board or even the entire panel may have to be replaced. Who is correct?

 A. A only

 B. B only

 C. Both A and B

 D. Neither A nor B

 Answer A is incorrect. Technician B is also correct. Check for simple and obvious faults first.

 Answer B is incorrect. Technician A is also correct. If an LED has quit working, it may be necessary to repair the circuit board or replace the entire control panel to restore the lighting.

 Answer C is correct. Both Technicians are correct.

 Answer D is incorrect. Neither Technician is incorrect.

52. Moving the heater control switch knob from HEAT to DEFROST causes a vehicle's A/C compressor to engage and the underhood electric fan to start. Which of the following could be the cause?

 A. A grounded compressor clutch circuit

 B. A faulty heater control switch

 C. A faulty fan circuit

 D. The A/C circuit is working correctly

 Answer A is incorrect. A grounded clutch circuit would keep the compressor engaged at all times.

 Answer B is incorrect. The heater switch is performing as it should.

 Answer C is incorrect. The fan circuit is operating as it should.

 Answer D is correct. The A/C system is designed to operate the compressor and the fan when the DEFROST switch position is selected.

53. Which of the following would happen if the A/C condenser becomes blocked with leaves and debris?

TASK G.3

A. The A/C system would blow too much cold air.

B. The high-side refrigerant pressure in the A/C system would rise.

C. The evaporator would frost up.

D. The condenser would frost up.

Answer A is incorrect. If anything, A/C performance would be limited.

Answer B is correct. Blockage of cooling air to the condenser would likely cause high-side pressure to rise.

Answer C is incorrect. The evaporator would receive less cold refrigerant

Answer D is incorrect. The condenser would get hotter.

54. A vehicle's serpentine belt squeals whenever the A/C is turned to ON. Technician A says the compressor may be locked up. Technician B says the serpentine belt may be loose. Who is correct?

TASK G.5

A. A only

B. B only

C. Both A and B

D. Neither A nor B

Answer A is incorrect. Technician B is also correct.

Answer B is incorrect. Technician A is also correct.

Answer C is correct. Both Technicians are correct. Either of these conditions could cause the A/C belt to squeal.

Answer D is incorrect. Neither Technician is incorrect.

55. With the A/C ON, water is seen blowing from the A/C vents when the blower motor is turned to HIGH. Which of the following is the most likely cause?

TASK G.6

A. A leaking A/C condenser

B. A leaking expansion valve

C. A plugged heater core

D. A plugged condensate drain

Answer A is incorrect. The condenser is under the hood.

Answer B is incorrect. The leak described would not occur due to the expansion valve.

Answer C is incorrect. This would not cause the issue described; however, a leaking heater core could.

Answer D is correct. If the drain is blocked with leaves or debris, condensation would build up in the blower box and be thrown by the blower fan out of the vents.

PREPARATION EXAM 3—ANSWER KEY

1.	C	20.	C	39.	B
2.	B	21.	D	40.	C
3.	C	22.	B	41.	D
4.	A	23.	C	42.	C
5.	B	24.	C	43.	B
6.	B	25.	D	44.	C
7.	D	26.	A	45.	B
8.	D	27.	A	46.	C
9.	D	28.	B	47.	A
10.	B	29.	B	48.	A
11.	D	30.	A	49.	C
12.	D	31.	B	50.	C
13.	C	32.	B	51.	C
14.	A	33.	A	52.	C
15.	C	34.	C	53.	B
16.	A	35.	A	54.	C
17.	C	36.	B	55.	D
18.	C	37.	C		
19.	A	38.	C		

PREPARATION EXAM 3—EXPLANATIONS

TASK A.4

1. An oil pan is to be removed and replaced. Which of the following is LEAST LIKELY to be used to seal the oil pan when it is reinstalled?

    A. a cork gasket
    B. a synthetic gasket
    C. silicone-based sealant
    D. RTV sealant

    Answer A is incorrect. Cork gaskets are commonly used.

    Answer B is incorrect. Synthetic gasket materials are common.

    Answer C is correct. Silicone sealant is least likely to be used to seal an oil pan. Silicone is more likely to be used for bathtub caulking rather than for oil pan gaskets.

    Answer D is incorrect. In some cases, RTV may be used in place of a gasket.

2. All of the following cooling system components should be periodically inspected and pressure tested EXCEPT for the:

TASK A.6

A. radiator.

B. evaporator.

C. heater hoses.

D. pressure cap.

Answer A is incorrect. The radiator should be tested.

Answer B is correct. The evaporator is an A/C component, not part of the cooling system. It is also located inside the blower motor box so is not easily "inspected."

Answer C is incorrect. Hoses should be inspected and tested.

Answer D is incorrect. The pressure cap should be periodically pressure tested.

3. A radiator hose is found to be soft and "squishy" when squeezed. Technician A says the hose has aged and may fail before too long. Technician B says that radiator hoses are sometimes uniquely shaped and replacement hoses will only fit on certain vehicle models. Who is correct?

TASK A.8

A. A only

B. B only

C. Both A and B

D. Neither A nor B

Answer A is incorrect. Technician B is also correct.

Answer B is incorrect. Technician A is also correct.

Answer C is correct. Both Technicians are correct.

Answer D is incorrect. Both Technicians are correct.

4. When a cooling system thermostat fails open it usually will:

A. not allow the engine to warm up.

B. cause the engine to overheat.

C. cause the cabin heater core to malfunction.

D. not allow the A/C to operate.

Answer A is correct. If faulty, a thermostat will usually fail to close. This protects the engine from overheating.

Answer B is incorrect. A faulty thermostat usually fails to close.

Answer C is incorrect. The heater core continues to work properly, even if the coolant fails to become hot.

Answer D is incorrect. The A/C will continue to function properly.

TASK A.12

5. On vehicles with a V-belt-driven fan, which of the following components drives the thermostatic clutch?

   A. The camshaft

   B. The crankshaft

   C. The balance shaft

   D. The intermediate shaft

Answer A is incorrect. The crankshaft drives the thermostatic clutch.

Answer B is correct. The crankshaft pulley drives the belt which drives the thermostatic clutch.

Answer C is incorrect. The balance shaft does not drive the thermostatic clutch.

Answer D is incorrect. An intermediate shaft does not drive the thermostatic clutch.

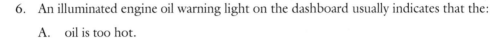

6. An illuminated engine oil warning light on the dashboard usually indicates that the:

   A. oil is too hot.

   B. oil pressure is too low.

   C. oil pressure relief valve is stuck closed.

   D. oil pressure switch is stuck open.

TASK A.13

Answer A is incorrect. Hot oil does not necessarily cause the oil light to illuminate.

Answer B is correct. The oil light switch closes the indicator light circuit when oil pressure drops to around 5 psi.

Answer C is incorrect. A stuck closed pressure relief valve would cause the oil pressure to be raised.

Answer D is incorrect. The oil light would not illuminate if the switch were stuck open.

TASK A.15

7. All of the following are routinely checked during a 12,000 mile maintenance inspection EXCEPT for the:

   A. air filter.

   B. oil level.

   C. tire pressures.

   D. cabin filter.

Answer A is incorrect. The air filter should be checked at routine intervals.

Answer B is incorrect. The oil level should be checked at routine intervals.

Answer C is incorrect. The tire pressures should be checked at routine intervals.

Answer D is correct. The cabin filter is not routinely checked at 12,000 mile intervals.

8. Which of the following is best for retrieving engine-related diagnostic trouble codes (DTCs) on an OBD II vehicle?

    A. A digital multimeter

    B. An analog voltmeter

    C. A lab-type oscilloscope

    D. A scan tool

TASK A.18

Answer A is incorrect. A DMM cannot read DTCs or freeze frame data.

Answer B is incorrect. An analog meter should not be used for solid state components.

Answer C is incorrect. A lab scope cannot retrieve DTCs or other data.

Answer D is correct. A scan tool is best for retrieving DTCs and stored freeze frame data in the PCM.

9. Shortly after the gas tank is filled, an MIL illuminates. Which of the following should the technician do first?

    A. Pull any trouble codes in the PCM.

    B. Take a look at freeze frame data in the PCM.

    C. Check under the hood for loose or faulty EVAP hoses.

    D. Check if the gas cap is loose.

TASK A.21

Answer A is incorrect. The most likely cause is a loose gas cap.

Answer B is incorrect. The gas cap should be checked first; it is commonly the problem and is easy to check.

Answer C is incorrect. Loose or faulty EVAP hoses are not the most likely cause of the MIL "ON" condition following a refueling event.

Answer D is correct. Check the gas cap first, as this is the most likely cause of the DTC.

10. Type "F" automatic transmission fluid (ATF) is used:

    A. for most automatic transmission vehicles.

    B. only for certain Ford® transmissions.

    C. only for Ferrari® transmissions.

    D. for vehicles made in Korea.

TASK B.2

Answer A is incorrect. Most auto transmissions use Mercon® or Dexron® ATF

Answer B is correct. Certain model year Ford® vehicles used type "F" ATF.

Answer C is incorrect. Type "F" ATF is used only in Ford® brand autos.

Answer D is incorrect. Korean vehicles do not use type "F" ATF.

11. An automatic transmission experiences a mild leak of trans fluid at the oil pan gasket. Technician A says that leak-stopping additive may be added to stop the leak. Technician B says to use RTV sealant at the leak area to stop the leak. Who is correct?

    A. A only

    B. B only

    C. Both A and B

    D. Neither A nor B

TASK B.3

Answer A is incorrect. The use of leak-stopping additives is not recommended.

Answer B is incorrect. Patching the leak is not a recommended repair.

Answer C is incorrect. Neither Technician is correct.

Answer D is correct. Neither Technician is correct. The pan bolts should be tightened, or the gasket should be replaced.

TASK B.4

12. A driveshaft universal joint (U-joint) is dry of lubricant and found to be loose. Which of the following is the proper repair?

A.   Grease the U-joint.

B.   Tighten the U-joint.

C.   Grease and tighten the U-joint.

D.   Replace the U-joint.

Answer A is incorrect. Greasing the U-joint would not cure the dry and rusting condition.

Answer B is incorrect. Tightening the U-joint would not cure the loose condition.

Answer C is incorrect. Neither greasing nor tightening will solve the problem, as these are not reliable repairs.

Answer D is correct. The old U-joint should be removed and replaced.

TASK B.6

13. A transmission mount is being inspected for wear. Technician A says to look for oil-soaked and swelled rubber between the mounting plates. Technician B says to look for separated, torn, or collapsed rubber between the mounting plates. Who is correct?

A.   A only

B.   B only

C.   Both A and B

D.   Neither A nor B

Answer A is incorrect. Technician B is also correct.

Answer B is incorrect. Technician A is also correct.

Answer C is correct. Both Technicians are correct.

Answer D is incorrect. Neither Technician is incorrect.

TASK C.3

14. A manual transmission vehicle is hard to shift. Technician A says that the shift linkage may be misaligned. Technician B says the hydraulic clutch reservoir may be overfilled. Who is correct?

A.   A only

B.   B only

C.   Both A and B

D.   Neither A nor B

Answer A is correct. Only Technician A is correct. The shift linkage may be bent or out of alignment.

Answer B is incorrect. An overfilled reservoir would not affect shift quality

Answer C is incorrect. Technician B is incorrect.

Answer D is incorrect. Technician A is correct.

15. A manual transmission's fluid level is being drained and refilled. Technician A says that some manual transmissions use the same type fluid as that used in engine crankcases. Technician B says that some manual transmissions use the same type fluid as that used in differentials. Who is correct?

TASK C.5

 A.   A only

 B.   B only

 C.   Both A and B

 D.   Neither A nor B

Answer A is incorrect. Technician B is also correct.

Answer B is incorrect. Technician A is also correct.

Answer C is correct. Both Technicians are correct.

Answer D is incorrect. Neither Technician is incorrect.

16. A driveshaft on a front wheel drive (FWD) vehicle with McPherson struts must be removed in order to replace an alternator. Technician A says the tie rod will first need to be disconnected from the steering knuckle. Technician B says that if a tie rod end is disconnected, a wheel alignment will be needed to correct the toe setting. Who is correct?

TASK C.7

 A.   A only

 B.   B only

 C.   Both A and B

 D.   Neither A nor B

Answer A is correct. Only Technician A is correct. The tie rod end needs to be disconnected in order to swing the steering knuckle out of the way so the drive axle can be slid out from the knuckle

Answer B is incorrect. The toe setting is not upset if the tie rod end is popped loose from the knuckle without changing its adjustment.

Answer C is incorrect. Technician B is incorrect.

Answer D is incorrect. Technician A is correct.

17. A few months after a front wheel bearing has been replaced in a rear wheel drive (RWD) pickup, the bearing is found to be making noise and the outer race it is found to be badly scored and wearing unevenly. Which of the following is the most likely cause of the premature bearing failure?

TASK C.9

 A.   An incorrect part number bearing was installed.

 B.   The axle nut was improperly torqued.

 C.   A small metal burr kept the bearing from properly seating.

 D.   The lugs nuts were improperly torqued.

Answer A is incorrect. The wrong part number would not likely be the cause of the uneven wear pattern.

Answer B is incorrect. An improperly torqued axle nut would not likely be the cause of a bearing failure.

Answer C is correct. A metal burr caused the bearing to seat improperly.

Answer D is incorrect. Incorrect lug nut torque would not be a likely cause for a bearing failure.

18. Transfer case fluid in a 4WD vehicle is to be drained and replaced. Technician A says that some transfer cases have the word "Drain" cast into the housing to indicate the drain plug location. Technician B says that some drain plugs can be removed by using a 3/8" extension on a ratchet wrench. Who is correct?

A. A only

B. B only

C. Both A and B

D. Neither A nor B

Answer A is incorrect. Technician B is also correct.

Answer B is incorrect. Technician A is also correct.

Answer C is correct. Both Technicians are correct. Some (for example, BorgWarner®) transfer cases have the word "Drain" cast into their housings. Drain plugs often can be removed/installed by using a 3/8" extension on a ratchet.

Answer D is incorrect. Both Technicians are correct.

19. All of the following statements about replacing a leaking transaxle half-shaft seal in a 4WD transverse engine vehicle are true EXCEPT the:

A. axle seal can be replaced without removing the half shaft.

B. half-shaft will have to be removed in order to replace the seal.

C. steering knuckle must be disconnected from the steering linkage.

D. large half-axle nut is easiest to loosen if done before the vehicle is raised from the shop floor.

Answer A is correct. The half shaft must be removed in order to install an axle seal.

Answer B is incorrect. The seal cannot be installed with the axle being removed first.

Answer C is incorrect. The steering knuckle will have to be disconnected in order to remove the half-axle.

Answer D is incorrect. The nut is torqued very tightly and loosening it with the wheel on the shop floor makes it easier to break it free.

20. Diagnostic trouble codes (DTCs) are to be retrieved on a vehicle. Technician A says that on some vehicles, simply turning the ignition ON-OFF-ON three times in succession will cause DTCs to display on the electronic odometer. Technician B says that some vehicles' DTCs can be read on the electronic climate control panel. Who is correct?

A. A only

B. B only

C. Both A and B

D. Neither A nor B

Answer A is incorrect. Technician B is also correct.

Answer B is incorrect. Technician A is also correct.

Answer C is correct. Both Technicians are correct. Some OBD II vehicles (Chrysler®) allow codes to be read by cycling the key. Some vehicle (GM®) allow fault codes to be read on the electronic climate control panel (ECCP).

Answer D is incorrect. Both Technicians are correct.

21. Power steering fluid is seen dripping from the power steering return hose where it connects to the PS reservoir. Which of the following is the most likely cause?

TASK D.4

    A.   The fluid is the wrong viscosity.
    B.   The fluid is the wrong type.
    C.   The banjo fitting washer is split.
    D.   The hose clamp needs to be tightened.

Answer A is incorrect. Incorrect viscosity fluid would not necessarily contribute to a leak.

Answer B is incorrect. The wrong type of fluid would not necessarily contribute to a leak.

Answer C is incorrect. Most PS systems do not use banjo fittings on the return line.

Answer D is correct. This is the most likely cause of fluid to be dripping; the hose clamp has come loose.

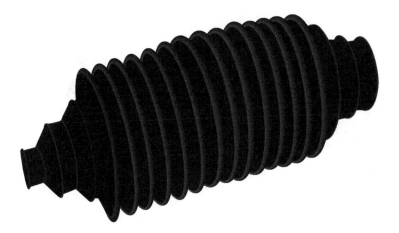

22. Which of the following is illustrated above?

TASK D.7

    A.   A CV joint boot
    B.   A rack and pinion boot
    C.   A strut boot
    D.   A shift lever boot

Answer A is incorrect. A CV joint boot is larger at one end.

Answer B is correct. This boot sometimes needs to be replaced on a power steering rack.

Answer C is incorrect. A strut dust boot looks different from this one pictured.

Answer D is incorrect. A shift lever boot is tapered in shape.

TASK D.16

23. Worn ball joints can best be checked for looseness and wear when the vehicle is:

A. fully loaded and bounced up and down.

B. empty and rocked from side to side.

C. raised slightly off the ground and each wheel is forced to move up and down.

D. raised slightly off the ground and each wheel is rotated.

Answer A is incorrect. This is a fictitious statement.

Answer B is incorrect. This is also a fictitious statement.

Answer C is correct. This is a standard safety inspection procedure.

Answer D is incorrect. Off the ground: Yes; rotate the wheels: No. This is a fictitious statement.

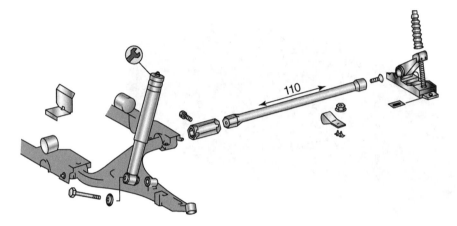

TASK D.21

24. What is the purpose of suspension part number 110 in the illustration above?

A. It prevents vehicle body sway.

B. It serves to dampen body roll.

C. It serves to soften the vehicle ride.

D. It serves to dampen body yaw.

Answer A is incorrect. It is not a sway bar.

Answer B is incorrect. Again, it is not a sway bar.

Answer C is correct. It is a torsion bar type spring.

Answer D is incorrect. Part no. 110 does not dampen fore and aft vehicle pitch.

25. The rear ride height on a live rear-axle pickup truck becomes too low (sags) when normal tools and equipment are being transported. Which of the following is the most likely cause?

TASK D.26

A. Weak struts

B. Weak shocks

C. Weak lateral/track bar

D. Weak leaf springs

Answer A is incorrect. Weak struts would affect comfort but not ride height.

Answer B is incorrect. Weak shocks struts would affect comfort but not ride height.

Answer C is incorrect. A faulty track bar would not affect ride height.

Answer D is correct. The springs are sagging and need to either be "assisted" with helper springs or replaced.

26. The rear struts are being inspected on a sports sedan and one is found to be leaking. Technician A says both rear struts should be replaced. Technician B says a front-end wheel alignment will be necessary after they are replaced. Who is correct?

TASK D.32

A. A only

B. B only

C. Both A and B

D. Neither A nor B

Answer A is correct. Only Technician A is correct. Struts on the same axle should be replaced in pairs.

Answer B is incorrect. Just a front-wheel alignment will not suffice; the rear-end alignment may have been upset, thus requiring a 4-wheel alignment.

Answer C is incorrect. Only Technician A is correct.

Answer D is incorrect. Only Technician B is incorrect.

27. A vehicle experiences excessive rear-end sway. Technician A says that the rear knuckle bushings may be worn. Technician B says that if a knuckle bushing goes bad, the entire knuckle will need to be replaced. Who is correct?

TASK D.36

A. A only

B. B only

C. Both A and B

D. Neither A nor B

Answer A is correct. Only Technician A is correct. A worn rear knuckle bushing could contribute to rear end sway.

Answer B is incorrect. A bushing can be pressed out of the knuckle and replaced.

Answer C is incorrect. Technician B is incorrect.

Answer D is incorrect. Technician A is correct.

TASK D.**40**

28. Technician A says that the object shown above, in the illustration is used for spreading a tire open when the tire is being patched. Technician B says that the object shown is for holding the steering wheel stationary and centered during a wheel alignment. Who is correct?

    A. A only

    B. B only

    C. Both A and B

    D. Neither A nor B

    Answer A is incorrect. This is not the intended purpose of the tool.

    Answer B is correct. Only Technician B is correct. The object shown is a steering wheel holding device.

    Answer C is incorrect. Only Technician B is correct.

    Answer D is incorrect. Only Technician A is incorrect.

TASK D.**41**

29. If a camber adjustment on a McPherson strut front end is to be performed, the:

    A. upper strut bearing may be moved fore or aft.

    B. upper strut bearing may be moved inward or outward.

    C. lower ball joint may be moved for or aft.

    D. lower ball joint may be moved inward or outward.

    Answer A is incorrect. Fore and aft movement affects caster.

    Answer B is correct. Camber adjustments require in and out movement. On these suspension systems, an aftermarket kit for this purpose is available and may be used.

    Answer C is incorrect. The lower ball joint is not adjustable.

    Answer D is incorrect. The lower ball joint is not adjustable.

30. A rear wheel toe adjustment:

    A. is not normally possible on a live rear-axle vehicle.

    B. may be performed at the tie rods of a live rear-axle vehicle.

    C. is not possible on a vehicle with an independently suspended rear end.

    D. may be performed on a vehicle with a solid rear axle.

TASK D.45

    Answer A is correct. A solid rear-axle vehicle does not normally have a toe adjustment provision.

    Answer B is incorrect. There are no tie rods on this type of vehicle.

    Answer C is incorrect. Tie rods do enable a toe adjustment on this type of suspension.

    Answer D is incorrect. A solid rear axle (a "live" rear axle) does not normally have a toe adjustment provision.

31. When driving straight, the steering wheel of a vehicle is found to be off center. Technician A says that resetting the steering wheel sensor will address the concern. Technician B says a wheel alignment will address the concern, but following the alignment, the steering wheel sensor will need to be reset. Who is correct?

TASK D.47

    A. A only

    B. B only

    C. Both A and B

    D. Neither A nor B

    Answer A is incorrect. A toe adjustment needs to be performed to center the off-center steering wheel.

    Answer B is correct. Only Technician B is correct. A wheel alignment must be performed followed by a steering wheel sensor reset.

    Answer C is incorrect. Only Technician B is correct.

    Answer D is incorrect. Only Technician A is incorrect.

TASK D.50

32. What is the purpose of the device shown above?

    A.   It is used when patching a tire.

    B.   It is used for checking tire pressures.

    C.   It is used for installing TPMS sensors.

    D.   It is used for checking tire tread depth.

    Answer A is incorrect. The item shown is not used for patching a tire per se, but it is used after the repair is finished.

    Answer B is correct. The item shown is a "stick type" tire pressure gauge.

    Answer C is incorrect. The item shown is not a TPMS tool.

    Answer D is incorrect. The item shown is not a tire tread-depth gauge.

TASK D.54

33. Which of the following is the LEAST LIKELY method to be used when dynamically balancing an out-of-balance tire and wheel assembly?

    A.   Fasten a weighted plate to the wheel rim.

    B.   Clamp a lead weight to the lip of a steel wheel rim.

    C.   Tape a lead weight to the inner rim of a custom wheel rim.

    D.   Clamp a zinc weight to the lip of the steel wheel rim.

    Answer A is correct. This is the LEAST LIKELY method of balancing an automotive wheel assembly. However, this method may be found used on tractors or other non-road vehicles.

    Answer B is incorrect. Using lead weights is the traditional means of tire balancing.

    Answer C is incorrect. Lead strips are routinely used on custom wheels to avoid damage to the rims.

    Answer D is incorrect. Zinc and steel weights are now used to replace environmentally harmful lead when balancing tires.

34. While attempting to replace a leaking wheel cylinder, a technician kinks the brake line near the wheel cylinder. Technician A says that new brake line steel tubing may be purchased, cut to length, double-flared, and correctly shaped as a replacement. Technician B says it's possible to purchase a straight brake line with flared ends and the correct fittings installed at each end as a replacement. Who is correct?

TASK E.3

A. A only
B. B only
C. Both A and B
D. Neither A nor B

Answer A is incorrect. Technician B is also correct.

Answer B is incorrect. Technician A is also correct.

Answer C is correct. Both Technicians are correct, although it's easier and quicker to follow Technician B's method of repair.

Answer D is incorrect. Both Technicians are correct.

35. Technician A says a hand-held vacuum pump may be used to help bleed a brake system. Technician B says a pressure bleeder requires two people to use. Who is correct?

TASK E.6

A. A only
B. B only
C. Both A and B
D. Neither A nor B

Answer A is correct. A hand vacuum pump may be used to extract air at the wheel brake's bleed screw.

Answer B is incorrect. A pressure bleeder requires only one person to use.

Answer C is incorrect. Technician B is incorrect.

Answer D is incorrect. Technician A is correct.

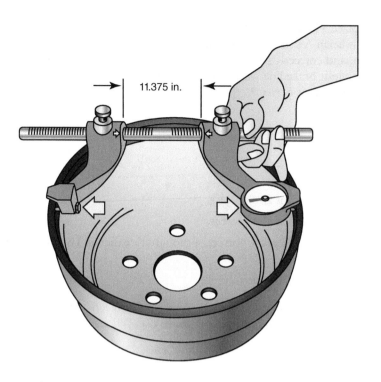

11.375 in.

TASK E.8

36. Technician A says the tool in the illustration above is being used to measure a brake drum's diameter so that the brake shoes may be pre-adjusted. Technician B says the drum is being checked for wear beyond its usable limit. Who is correct?

A.   A only

B.   B only

C.   Both A and B

D.   Neither A nor B

Answer A is incorrect. A different gauge is used for this purpose.

Answer B is correct. Only Technician B is correct. The drum micrometer shown is used for checking the degree of brake drum wear; it can also be used to check for warpage.

Answer C is incorrect. Only Technician B is correct.

Answer D is incorrect. Only Technician A is incorrect.

TASK E.12

37. Which of the following tools should be used when removing a brake line fitting from a wheel cylinder?

A.   An open-end wrench

B.   A box-end wrench

C.   A flare wrench

D.   A combination wrench

Answer A is incorrect. An open-end wrench could round off the soft metal brake line fitting.

Answer B is incorrect. A box-end wrench could not be placed around the brake line fitting.

Answer C is correct. A flare wrench has five sides with an opening which enables it to be placed on the brake line fitting.

Answer D is incorrect. A combination wrench is both an open-end and a box-end wrench. (See explanations A and B above.)

38. A vehicle is to be raised on a lift to have its parking brake system inspected for proper operation. Technician A says that, on a drum-type parking brake system, to inspect the parking brake cables, equalizer, and self-adjusting linkage. Technician B says a caliper-type parking brake system should have its cables, equalizer, cams, and screw mechanisms inspected. Who is correct?

TASK E.15

   A. A only
   B. B only
   C. Both A and B
   D. Neither A nor B

   Answer A is incorrect. Technician B is also correct

   Answer B is incorrect. Technician A is also correct.

   Answer C is correct. Both Technicians are correct about parking brake adjustments.

   Answer D is incorrect. Neither Technician is incorrect.

39. It is important to torque wheel lug nuts or bolts to OEM specs, and in the correct sequence, in order to help prevent:

TASK E.16

   A. brake piston binding.
   B. disc warpage.
   C. brake squeal.
   D. cracking the drums.

   Answer A is incorrect. Piston binding would not occur from improper tightening.

   Answer B is correct. Using the incorrect torque and tightening sequence of the lug bolts/ nuts could cause rotor warpage and a subsequent brake pedal pulsation condition.

   Answer C is incorrect. Brake squeal is not likely to be caused by an incorrect torque or tightening sequence.

   Answer D is incorrect. The drums would not likely crack.

40. A vehicle's brake calipers have experienced severe rust and corrosion both from exposure to a salt water environment and from non-use. Technician A says the calipers should be rebuilt. Technician B says the calipers should be replaced. Who is correct?

TASK E.19

   A. A only
   B. B only
   C. Both A and B
   D. Neither A nor B

   Answer A is incorrect. Technician B is also correct.

   Answer B is incorrect. Technician A is also correct.

   Answer C is correct. Both Technicians are correct. It's a judgment call, but based on the labor time involved in rebuilding the calipers combined with the risk of comebacks, it's probably more economical to exchange them for remanufactured calipers.

   Answer D is incorrect. Neither Technician is incorrect.

TASK E.23

41. Warped rotors are being replaced with new ones on a vehicle. Once installed, the new rotors will need to be:

A. measured for runout.

B. checked for warpage.

C. sanded to remove residual rust.

D. burnished, or "burned in."

Answer A is incorrect. New or rebuilt rotors should be true and free of runout.

Answer B is incorrect. New or remanufactured rotors should be true and free of warpage.

Answer C is incorrect. New rotors come coated with preservative and/or wrapped in sealed packaging.

Answer D is correct. This is an important step not to be overlooked. The burnishing process may vary depending on the manufacturer, but it involves repeated brake applications to transfer brake pad material onto the rotors for more effective braking.

TASK E.24

42. What happens if off-car brake drum machining is performed too fast on a brake lathe, with only a rough cut and no fine-finish cut made?

A. Once in use, the brakes are likely to make a squealing noise.

B. Once in use, the brake shoes will overheat.

C. The shoes will "click" as they attempt to follow spiral grooves in the drum.

D. The brakes are likely to make a humming noise.

Answer A is incorrect. Machining in the fashion described would not cause brake squeal.

Answer B is incorrect. Under normal use, overheating is not likely to occur.

Answer C is correct. The shoes will move much like a phonograph arm follows the grooves in a record; that is, they will "click" after they move sideways and when they "jump back" to re-center themselves.

Answer D is incorrect. Humming would not occur.

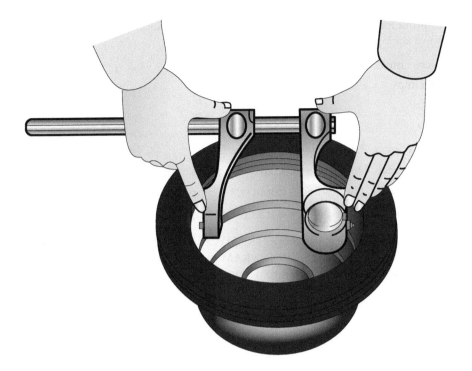

43.　What action is taking place in the illustration above?

　　A.　A brake drum is being measured for wear.

　　B.　A parking brake drum's inside diameter is being measured.

　　C.　A brake rotor's circumference is being measured.

　　D.　A brake rotor is being checked for warpage.

TASK E.26

Answer A is incorrect. A brake drum is not pictured.

Answer B is correct. The inside diameter of a parking brake drum (inside part of a rotor) is being measured

Answer C is incorrect. The circumference is not being measured.

Answer D is incorrect. Rotor warpage is not being checked.

44.　A power brake booster is being checked for proper operation. Technician A says if when the engine is run and then turned off, the brake pedal is pumped until the pedal feels harder and higher, and the engine is re-started, the brake pedal should drop slightly upon engine restart. Technician B says to first start the engine, push and hold the brake pedal down, turn off the engine and pump the pedal to see if the brake pedal feels harder and rises. Who is correct?

TASK E.32

　　A.　A only

　　B.　B only

　　C.　Both A and B

　　D.　Neither A nor B

Answer A is incorrect. Technician B is also correct.

Answer B is incorrect. Technician A is also correct.

Answer C is correct. Both Technicians are correct. The statements are two sides of the same coin; both are valid tests of the brake booster and its check valve.

Answer D is incorrect. Neither Technician is incorrect.

45. While measuring for current in a circuit, a DMM reads "0.035". Which of the following is another way a stating the DMM's reading?

A. 35 micro amps

B. 35 milliamps

C. 3.5 amps

D. 35 amps

Answer A is incorrect. The reading is not in microamps.

Answer B is correct. The meter reads thirty-five thousandths of an amp, or 35 milliamps.

Answer C is incorrect. The reading is not in whole amps.

Answer D is incorrect. The reading is not in whole amps.

46. Technician A says that keep alive memory (KAM) can be reset by disconnecting the battery for about 5 minutes. Technician B says that KAM memory functions for the radio, seat position, clock and other settings prior to disconnecting the battery. Who is correct?

A. A only

B. B only

C. Both A and B

D. Neither A nor B

Answer A is incorrect. Technician B is also correct.

Answer B is incorrect. Technician A is also correct

Answer C is correct. Both Technicians are correct. Often, any stored KAM in the PCM is erased (reset) when 12-volt battery power is lost for a period of time. As a customer convenience, the radio station presets, seat position, clock setting etc. should be recorded before KAM is lost, and then reprogrammed by the technician once the battery is reconnected.

Answer D is incorrect. Neither Technician is incorrect.

47. Technician A says a specially designed wire brush is available to clean the terminals of a side-terminal battery. Technician B says that baking powder can be used to neutralize corrosion on battery terminals. Who is correct?

A. A only

B. B only

C. Both A and B

D. Neither A nor B

Answer A is correct. Only Technician A is correct. A special circular wire brush is useful for cleaning these terminals, especially when they are in tight locations.

Answer B is incorrect. Baking soda should be used to clean corrosion, not baking powder.

Answer C is incorrect. Only Technician A is correct.

Answer D is incorrect. Only Technician B is incorrect.

48. A vehicle with a fully charged battery does not make a "click" sound or crank the engine when the ignition key is turned to the START position. Which of the following could be the cause?

TASK F.12

   A. A faulty neutral-safety switch
   B. A faulty starter motor
   C. A faulty automatic transmission clutch switch
   D. A faulty battery

   Answer A is correct. A faulty neutral-safety switch would prevent the starter solenoid from getting power to "click" (activate).

   Answer B is incorrect. The starter is likely OK; the control circuit is faulty.

   Answer C is incorrect. A vehicle with an automatic transmission does not have a clutch or clutch safety switch.

   Answer D is incorrect. The battery is fully charged and OK.

49. A charging system output test is to be performed by placing a high-amperage "amps clamp" on the heavy lead from the battery to the alternator. The clamp's "+" sign is pointing toward the battery. If the charging system is loaded and working properly, the digital reading would show:

TASK F.14

   A. high-amperage positive output from the alternator.
   B. low-amperage positive output from the alternator.
   C. high-amperage negative (minus) output from the alternator.
   D. a high-amperage reading to the starter.

   Answer A is incorrect. The clamp is reversed and will read backwards current flow.

   Answer B is incorrect. The clamp will read high-amperage flow.

   Answer C is correct. The clamp is installed backwards, so the reading will be reversed.

   Answer D is incorrect. Starter draw is not being measured with this alternator load test.

50. A rebuilt alternator is being installed in a vehicle. Which of the following could likely occur if a crowbar is used to tension the alternator when tightening the mounting bracket and serpentine drive belt?

TASK F.16

   A. The alternator fields could be damaged.
   B. The alternator rotor may be damaged.
   C. The mounting ear(s) could be broken off.
   D. The drive belt could be snapped.

   Answer A is incorrect. The fields are not likely to be damaged.

   Answer B is incorrect. The rotor is not likely to be damaged.

   Answer C is correct. Breaking off the mounting ears is a common mishap when prying on an alternator.

   Answer D is incorrect. Snapping the belt is highly unlikely.

TASK F.19

51. A vehicle's plastic headlight lenses are clouded enough to cause the vehicle to fail a safety inspection. Technician A says that the lenses can restored by using a fine grade of rubbing compound or toothpaste. Technician B says the lenses can be restored using a restoration kit containing fine grit sandpaper and a special liquid. Who is correct?

   A. A only
   B. B only
   C. Both A and B
   D. Neither A nor B

   Answer A is incorrect. Technician B is also correct.

   Answer B is incorrect. Technician A is also correct.

   Answer C is correct. Both Technicians are correct. Both of these methods have been successfully used to help restore clouded plastic headlight lenses or covers.

   Answer D is incorrect. Neither Technician is incorrect.

TASK F.21

52. A vehicle's horn continues to blow whenever the ignition is switched to the ON position. Which of the following could be the cause?

   A. A faulty ignition switch
   B. A faulty horn
   C. A faulty horn relay
   D. An open in the horn circuit

   Answer A is incorrect. A faulty ignition switch would cause the horn to activate.

   Answer B is incorrect. The horn obviously works.

   Answer C is correct. The horn relay is likely sticking closed.

   Answer D is incorrect. An open circuit would prevent the relay from being activated.

TASK G.2

53. Signs of oil seepage are seen at the A/C connecting block on a vehicle's firewall. Which of the following is the most likely cause?

   A. A faulty connecting block
   B. A faulty "O" ring at the connection
   C. A broken fitting
   D. A broken connection at the compressor

   Answer A is incorrect. The block is not likely to break.

   Answer B is correct. An "O" ring is more likely to dry out and leak.

   Answer C is incorrect. The fitting is not likely to break.

   Answer D is incorrect. The compressor is not located on the firewall.

TASK G.3

54. A vehicle's A/C condenser seems too hot, and overall A/C performance is poor. Which of the following is the most likely cause?

   A. The A/C clutch is weak and slipping.
   B. The evaporator is leaking refrigerant.
   C. The condenser's airflow is blocked by debris.
   D. The compressor is worn out.

   Answer A is incorrect. A weak A/C clutch would not cause the condenser to be overheated.

   Answer B is incorrect. Low refrigerant would not cause the condenser to be too hot.

   Answer C is correct. Insufficient airflow will keep the condenser from properly cooling.

   Answer D is incorrect. If anything, the condenser would be too cool.

55. A vehicle's cabin air filter is due for inspection. If found to be dirty, it would normally be:

    A. blown free of debris with low pressure shop air and reinstalled.

    B. washed with mild dish soap, rinsed, air dried and reinstalled.

    C. cleaned of debris with a shop vacuum cleaner and reinstalled.

    D. replaced.

TASK G.4

Answer A is incorrect. Blowing away debris and reinstalling the cabin filter is not a correct procedure.

Answer B is incorrect. Washing a cabin filter is not normally done.

Answer C is incorrect. Cleaning a cabin filter with a shop vac would not be the correct procedure.

Answer D is correct. The filter is normally discarded and replaced at the OEM-specified mileage interval.

PREPARATION EXAM 4—ANSWER KEY

1.	C	**20.**	A	**39.**	D
2.	B	**21.**	A	**40.**	C
3.	C	**22.**	A	**41.**	B
4.	B	**23.**	D	**42.**	A
5.	B	**24.**	C	**43.**	C
6.	B	**25.**	A	**44.**	B
7.	D	**26.**	B	**45.**	D
8.	A	**27.**	C	**46.**	B
9.	C	**28.**	A	**47.**	D
10.	A	**29.**	C	**48.**	C
11.	C	**30.**	A	**49.**	D
12.	C	**31.**	D	**50.**	D
13.	C	**32.**	A	**51.**	D
14.	B	**33.**	C	**52.**	D
15.	C	**34.**	B	**53.**	A
16.	B	**35.**	A	**54.**	A
17.	B	**36.**	C	**55.**	A
18.	B	**37.**	B		
19.	C	**38.**	D		

PREPARATION EXAM 4—EXPLANATIONS

TASK A.2

1. A vehicle engine is leaking oil from its bell housing. Technician A says the PCV valve should be inspected for being clogged. Technician B says the rear oil seal probably needs to be replaced. Who is correct?

   A. A only

   B. B only

   C. Both A and B

   D. Neither A nor B

   Answer A is incorrect. B is also correct.

   Answer B is incorrect. A is also correct.

   Answer C is correct. Both Technicians are correct. A clogged PCV valve could cause a buildup of crankcase blow-by pressure which would cause the rear seal to be damaged and leak oil. The rear seal may therefore need to be replaced along with the PCV valve.

   Answer D is incorrect. Both Technicians are correct.

TASK A.4

2. An engine oil pan has been removed in order to replace a leaking gasket. At the same time the oil pan should be inspected for any of the following EXCEPT:

   A. dents.

   B. scratches.

   C. cracks.

   D. chips.

   Answer A is incorrect. The pan should be inspected for dents.

   Answer B is correct. It would be very unlikely for an oil pan to have a deep enough scratch to leak.

   Answer C is incorrect. The pan should be inspected for cracks.

   Answer D is incorrect. The pan should be inspected for chips.

3. Modern car engines generally require oil changes at:

   A. 2,500 mile intervals.
   B. 3,000 mile intervals.
   C. 5,000 mile intervals.
   D. 10,000 mile intervals.

TASK A.5

Answer A is incorrect. 2,500 miles is too frequent.

Answer B is incorrect. Under normal vehicle use, 3,000 mile intervals is also too frequent.

Answer C is correct. Generally speaking, many OEMs recommend 5,000 (or 6,000) mile oil change intervals

Answer D is incorrect. For most vehicles using organic (non-synthetic) motor oil, 10,000 miles is too much. Such an extended oil change interval would not be recommended by OEMs.

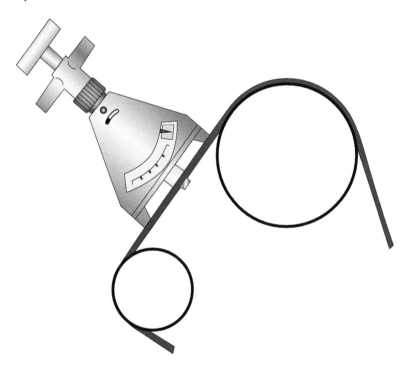

4. The tool in the illustration above is being used to check a:

   A. drive belt for wear.
   B. drive belt for correct tension.
   C. pulley for alignment.
   D. tensioner for alignment.

TASK A.7

Answer A is incorrect. The tool shown is a belt tension gauge.

Answer B is correct. The tool shown is only one of a number of different styles of belt tension gauges.

Answer C is incorrect. The tool shown is a belt tension gauge.

Answer D is incorrect. The tool shown is a belt tension gauge.

TASK A.10

5. A cooling system is to be drained. Technician A says that in order to get all impurities out of the cooling system, it should be drained while hot. Technician B says in order to remove a radiator cap, it should be pushed down and turned counterclockwise. Who is correct?

A. A only

B. B only

C. Both A and B

D. Neither A nor B

Answer A is incorrect. Only Technician B is correct. The cooling system should only be opened and drained when COLD.

Answer B is correct. Only Technician B is correct. The statement correctly describes how to remove a radiator cap.

Answer C is incorrect. Only Technician B is correct

Answer D is incorrect. Only Technician A is incorrect.

TASK A.12

6. Which of the following normally drives the thermostatic clutch on vehicles with a belt-driven fan?

A. The camshaft

B. The crankshaft

C. The balance shaft

D. The intermediate shaft

Answer A is incorrect. The camshaft does not usually have a pulley to drive a thermostatic clutch.

Answer B is correct. The crankshaft pulley drives the thermostatic clutch with integrated fan assembly.

Answer C is incorrect. The balance shaft does not have a pulley to drive a thermostatic clutch.

Answer D is incorrect. The intermediate shaft is part of the steering system, not part of the engine.

TASK A.15

7. A routine 5,000 mile interval inspection includes checking the:

A. transmission filter.

B. power steering fluid filter.

C. cabin filter.

D. air filter.

Answer A is incorrect. The transmission is not routinely inspected at every service interval.

Answer B is incorrect. The power steering filter is not routinely inspected at every service interval.

Answer C is incorrect. The cabin filter is not routinely inspected at every service interval.

Answer D is correct. The air filter should be inspected at each service interval to make certain it has not been contaminated or blocked by debris.

8. The trouble codes on an OBD I Ford® are to be retrieved. Technician A says to use the self-test connector located under the hood on the driver's side of the vehicle. Technician B says to insert a jumper wire in the self-test connector and use a digital multimeter to count the needle swings and read the codes. Who is correct?

TASK A.18

   A.   A only
   B.   B only
   C.   Both A and B
   D.   Neither A nor B

Answer A is correct. Only Technician A is correct. The self-test connector is a "house shaped" connector under the hood on the driver's side.

Answer B is incorrect. A digital meter does not have a needle, as opposed to an analog meter. On these vehicles, fault codes can also be read by inserting the jumper wire and watching the "check engine" light blink.

Answer C is incorrect. Only Technician A is correct

Answer D is incorrect. Only Technician B is incorrect.

9. A 1996 model OBD II vehicle registered and driven in the "snowbelt" region of the country fails an emissions test. Which of the following is the most likely cause?

TASK A.20

   A.   A faulty fuel pump
   B.   A cracked EVAP canister
   C.   A rusted fuel filler neck
   D.   A leaking fuel rail

Answer A is incorrect. A faulty fuel pump would not cause an EVAP leak.

Answer B is incorrect. A cracked EVAP canister is not the most likely cause.

Answer C is correct. The ice-melting salt used in "snowbelt" areas of the country commonly rusts the exposed fuel filler pipe.

Answer D is incorrect. A leaking fuel rail is not the most likely cause.

10. The volatility of gasoline in a vehicle's tank is to be checked. Technician A says that testing the fuel's Reid Vapor Pressure will reveal the fuel's volatility. Technician B says that higher fuel volatility fuel is sold during the summer months. Who is correct?

TASK B.2

   A.   A only
   B.   B only
   C.   Both A and B
   D.   Neither A nor B

Answer A is correct. Only Technician A is correct. RVP testing reveals the fuel's volatility.

Answer B is incorrect. Greater fuel volatility is required during colder seasons when fuel evaporates less easily.

Answer C is incorrect. Only Technician A is correct.

Answer D is incorrect. Only Technician B is incorrect.

TASK B.5

11. Technician A says that a leaking transmission line may not be apparent because the fluid (ATF) could burn off before it drips from the leak to the shop floor. Technician B says that a transmission line leak would likely appear as red colored ATF. Who is correct?

    A. A only
    B. B only
    C. Both A and B
    D. Neither A nor B

    Answer A is incorrect. Technician B is also correct.

    Answer B is incorrect. Technician A is also correct.

    Answer C is correct. Both Technicians are correct. ATF may collect and burn on the exhaust before dripping to the driveway or shop floor. If it leaks, it would normally appear as red colored fluid.

    Answer D is incorrect. Neither Technician is incorrect.

TASK B.6

12. Technician A says that powertrain mounts can be checked for defects by revving the engine in gear with the brakes firmly applied and watching how much the engine lifts up as it applies torque to the drive line. Technician B says that an engine mount can be inspected by jacking up the engine slightly and looking for an air gap where the mount's cushioning rubber should be. Who is correct?

    A. A only
    B. B only
    C. Both A and B
    D. Neither A nor B

    Answer A is incorrect. Technician B is also correct.

    Answer B is incorrect. Technician A is also correct.

    Answer C is correct. Both Technicians are correct. Both of the methods described may be used to check for defective engine mounts.

    Answer D is incorrect. Neither Technician is incorrect.

TASK B.7

13. After a transmission pan has been removed to replace the transmission filter, a large magnet is found in the pan with gooey grey matter adhering to it. Which of the following statements best describes why the magnet is in the transmission pan?

    A. The magnet was inadvertently left in the transmission pan during transmission assembly.
    B. The magnet is in the pan to collect dirt which finds its way into the transmission fluid.
    C. The magnet is in the pan to collect fine metal wear particles in the transmission fluid.
    D. The magnet is in the pan to collect large metal debris in the event of a transmission failure.

    Answer A is incorrect. The magnet is intended to be there.

    Answer B is incorrect. Dirt does not adhere to a magnet.

    Answer C is correct. The magnet's purpose is to retain fine metal wear particles and keep them out of the fluid.

    Answer D is incorrect. Heavy metal debris in the fluid can only mean a transmission that is soon to self-destruct.

14. The column shift mechanism on an older pickup truck with a manual transmission is sloppy and won't allow the transmission to go into all gears. Which of the following is the most likely cause?

    A. The shaft attached to the column shift lever is bent.
    B. The column shift lever shaft retaining bolts behind the dashboard have become loose.

TASK C.3

    C. The column shift lever shaft is broken.
    D. The shift cable is kinked.

    Answer A is incorrect. A bent shaft is not the most likely cause.

    Answer B is correct. This is a common complaint on certain makes/models of vehicles.

    Answer C is incorrect. A broken shift lever shaft is very unlikely.

    Answer D is incorrect. A cable would not be used for a manual transmission.

15. A manual transmission driveshaft seal is to be replaced. Technician A says to unbolt the driveshaft and remove the U-joint yoke to get at the seal. Technician B says that if there is too much play or slop in the U-joint yoke, a new seal will likely leak shortly after it's installed. Who is correct?

    A. A only
    B. B only

TASK C.4

    C. Both A and B
    D. Neither A nor B

    Answer A is incorrect. Technician B is also correct.

    Answer B is incorrect. Technician A is also correct.

    Answer C is correct. Both Technicians are correct about the driveshaft seal.

    Answer D is incorrect. Neither Technician is incorrect.

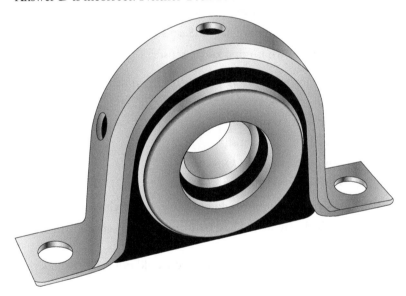

16. Which of the following driveshaft parts is pictured in the illustration above?

    A. Surface bearing
    B. Center support bearing
    C. Torrington bearing

TASK C.8

    D. Thrust bearing

    Answer A is incorrect. The image is not of a surface (rod) bearing.

    Answer B is correct. The bearing shown is used at the center of a two-piece driveshaft.

    Answer C is incorrect. The bearing shown is not a Torrington (end) bearing.

    Answer D is incorrect. The bearing shown is not a thrust bearing.

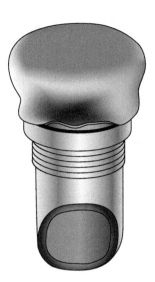

TASK C.13

17. Which of the following parts is illustrated above?

A. A tire valve

B. An axle housing vent

C. A TPMS sensor

D. A knock sensor

Answer A is incorrect. The part shown is not a tire valve.

Answer B is correct. The part shown is a vent for an axle/differential housing.

Answer C is incorrect. The part shown is not a TPMS sensor.

Answer D is incorrect. The part shown is not a knock sensor.

TASK C.16

18. With the engine running, the most obvious sign of a bad U-joint is a loud and hard clunk when:

A. coasting with the transmission in neutral.

B. placing the vehicle in DRIVE at a standstill.

C. trailing the throttle from a medium road speed.

D. shifting to neutral while at a standstill.

Answer A is incorrect. Coasting in neutral would not cause a loud clunk.

Answer B is correct. Driveshaft play would best be heard when placing a load on the drivetrain at a standstill.

Answer C is incorrect. Trailing the throttle art this speed would not cause a clunk that would be heard.

Answer D is incorrect. Shifting to neutral would not cause a loud clunk to be heard.

19. Technician A says that for cars and vans, staying within a 3% diameter change of the OEM tie size is acceptable. Technician B says that for pickups and sport utility vehicles (SUVs), room for up to a 15% oversize tire is generally provided for when the vehicle is designed. Who is correct?

TASK C.20

   A.   A only
   B.   B only
   C.   Both A and B
   D.   Neither A nor B

   Answer A is incorrect. Technician B is also correct.

   Answer B is incorrect. Technician A is also correct.

   Answer C is correct. Both Technicians are correct.

   Answer D is incorrect. Neither Technician is incorrect.

20. The power steering fluid level is to be checked on a vehicle. Technician A says that on some cars the power steering fluid level can only be checked accurately after the engine has run for a brief period. Technician B says that on some cars a dipstick is used for checking the PS fluid only when it is "cold." Who is correct?

TASK D.2

   A.   A only
   B.   B only
   C.   Both A and B
   D.   Neither A nor B

   Answer A is correct. Only Technician A is correct. Some vehicles require the fluid to be warmed up when checked.

   Answer B is incorrect. Most PS fluid dipsticks generally have both "cold" and "warm" gradations (indicator lines).

   Answer C is incorrect. Only Technician A is correct

   Answer D is incorrect. Only Technician B is incorrect.

21. Abnormal noises made by a vane-type power steering pump may include any of the following EXCEPT a:

TASK D.4

   A.   low frequency shudder.
   B.   low frequency moan.
   C.   high frequency whine.
   D.   high frequency hiss.

   Answer A is correct. A vane-type PS pump may cause a mechanical shudder felt in the steering wheel, but this is not associated with normally heard (audible) noise.

   Answer B is incorrect. A moan may be heard when low on PS fluid.

   Answer C is incorrect. A whine may indicate a PS pump problem.

   Answer D is incorrect. A hiss may be heard caused by fluid flowing across a valve in the PS pump.

TASK D.10

22. Which of the following is the correct procedure for adjusting toe on a vehicle?

    A.   Loosen and turn the adjusting sleeves with the proper tie rod sleeve tool.

    B.   Loosen and turn the castellated nuts on the tie rod ends.

    C.   Rotate the left tie rod end until the correct toe is achieved.

    D.   Rotate the right tie rod end until the correct toe is achieved.

Answer A is correct. The procedure described is how toe is to be adjusted.

Answer B is incorrect. The castellated nut secures the tie rod end to the steering knuckle.

Answer C is incorrect. The tie rod itself is not rotated.

Answer D is incorrect. Tie rods are not rotated.

TASK D.11

23. If the part on an import vehicle as illustrated above is found to be leaking, which of the following describes how vehicle handling would be affected?

    A.   The vehicle will continue to bounce up and down after hitting bumps.

    B.   The vehicle will not track straight when the steering wheel is not held.

    C.   The vehicle will lean excessively during turns.

    D.   The vehicle will experience steering wheel jerkiness when hitting bumps.

Answer A is incorrect. The part shown is not shock absorber.

Answer B is incorrect. The part shown does not affect toe.

Answer C is incorrect. The part shown is not a sway bar.

Answer D is correct. The part is a Volkswagen steering damper. The damper prevents road shock from being transferred to the steering linkage.

24. Upon inspection, a vehicle's front ride height is low and the front coil springs show signs of coil contact. Which of the following should be done first?

    A. Replace the jounce bumpers.

    B. Replace the shocks.

    C. Replace the springs.

    D. Replace the tires.

TASK D.19

    Answer A is incorrect. The bumpers are not the root cause of the problem.

    Answer B is incorrect. The shocks, while perhaps a contributing factor, are also not the root cause of the problem.

    Answer C is correct. Because the springs show signs of weakness due to low ride height, they should be replaced.

    Answer D is incorrect. If the tires are the correct size for the vehicle, replacing them would not fix the problem.

25. A non-power steering vehicle with McPherson strut suspension is very difficult to turn to the left or right. Which of the following could be the cause?

    A. Faulty strut bearings

    B. Worn steering damper

    C. Worn struts

    D. Faulty tie rods

TASK D.24

    Answer A is correct. Faulty strut bearings could make turning the front wheels difficult, especially without power steering.

    Answer B is incorrect. A worn damper would not make steering difficult.

    Answer C is incorrect. Worn struts would affect the ride, but not the steering effort.

    Answer D is incorrect. Faulty tie rods could make steering sloppy, but would not make steering difficult.

26. A rear leaf spring equipped vehicle seems to sag on the right rear side. Technician A says that a leaf in the left side leaf spring may have snapped. Technician B says that the right side spring shackle may be damaged. Who is correct?

    A. A only

    B. B only

    C. Both A and B

    D. Neither A nor B

TASK D.30

    Answer A is incorrect. The right side likely has a snapped leaf, not the left.

    Answer B is correct. Only Technician B is correct. A faulty spring shackle could cause a sagging condition.

    Answer C is incorrect. Only Technician B is correct.

    Answer D is incorrect. Only Technician A is incorrect.

TASK D.33

27. A front wheel drive van with non-independent rear suspension shows severe rust and corrosion of the "U" shaped rear axle. Technician A says that such corrosion may weaken the axle and make it vulnerable to torsional stress. Technician B says that such a condition could lead to cracks or complete breakage of the axle. Who is correct?

   A. A only
   B. B only
   C. Both A and B
   D. Neither A nor B

Answer A is incorrect. Technician B is also correct.

Answer B is incorrect. Technician A is also correct

Answer C is correct. Both Technicians are correct. A major OEM issued a recall due to these rusting problems, testifying to the importance of performing a close inspection of vehicles with this type of rear axle.

Answer D is incorrect. Neither Technician is incorrect.

TASK D.36

28. The spindle assembly at the rear of the vehicle shows signs of severe wear. Which of the following could be the cause?

   A. Dry wheel bearings
   B. Incorrect rear camber
   C. Incorrect toe setting
   D. Worn rear tires

Answer A is correct. The spindle would show wear if the outer wheel bearing races should spin on the spindle.

Answer B is incorrect. Incorrect camber would not affect the spindle.

Answer C is incorrect. Incorrect toe would not affect the spindle.

Answer D is incorrect. Worn tires would not affect the spindle.

TASK D.38

29. A growling noise is heard at the rear of an FWD independently suspended rear-end vehicle. Technician A says it could be caused by a bent spindle. Technician B says it could be caused by a faulty wheel bearing. Who is correct?

   A. A only
   B. B only
   C. Both A and B
   D. Neither A nor B

Answer A is incorrect. Technician B is also correct.

Answer B is incorrect. Technician A is also correct.

Answer C is correct. Both Technicians are correct. A bent spindle would cause a bearing to fail. In turn, a failed bearing would make a growling noise.

Answer D is incorrect. Neither Technician is incorrect.

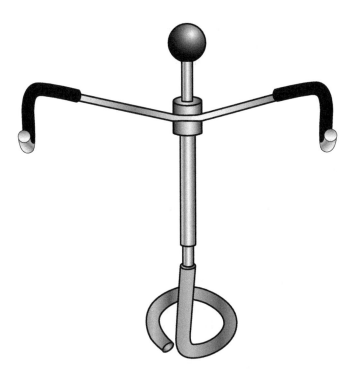

30. The object shown in the illustration above is used to:

A. secure a steering wheel in position during a wheel alignment.

B. hold the steering wheel during brake service.

C. support a brake caliper once it has been unbolted from a wheel assembly.

D. hold a seat assembly off the floor after it has been removed from a vehicle.

TASK D.**44**

Answer A is correct. The tool shown secures the steering wheel in a straight ahead position during a wheel alignment.

Answer B is incorrect. The tool is not used to hold the steering wheel during brake service.

Answer C is incorrect. This tool is not for holding a brake caliper during service.

Answer D is incorrect. The tool is not used to support a seat assembly

31. A new set of tires is to be installed on a vehicle. Technician A says that lower speed rated tires may be installed if the driver never drives the vehicle off his privately owned farm. Technician B says that the inflation pressure indicated on the tire sidewall is the inflation pressure which should be used. Who is correct?

TASK D.**50**

A. A only

B. B only

C. Both A and B

D. Neither A nor B

Answer A is incorrect. Technician B is also incorrect.

Answer B is incorrect. Technician A is also incorrect.

Answer C is incorrect. Neither Technician is correct. Lower speed rated tires should never be substituted for higher ones. The correct tire inflation pressure is indicated on a placard on the vehicle's driver-side door or door jamb.

Answer D is correct. Neither Technician is correct.

TASK D.54

32. Technician A says a tire puncture should be repaired before the tire assembly is balanced. Technician B says a tire may not need to be rebalanced if only a plug-type tire repair is performed and the wheel weights remained on the rim. Who is correct?

 A. A only

 B. B only

 C. Both A and B

 D. Neither A nor B

Answer A is correct. Only Technician A is correct. Following a repair, the tire should be rebalanced.

Answer B is incorrect. A correct tire repair means breaking down the tire from the rim. This means the wheel weight(s) must first be removed, and the wheel must subsequently be rebalanced after the repair.

Answer C is incorrect. Only Technician A is correct.

Answer D is incorrect. Only Technician B is incorrect.

TASK E.2

33. Following replacement of brake pads on a four-wheel disc brake equipped vehicle, the master cylinder reservoir is found to be overfilled and brake fluid is dripping from it. Which of the following is the most likely cause?

 A. The master cylinder port is clogged and causing brake fluid to back up.

 B. The reservoir is cracked and leaking brake fluid.

 C. The reservoir was not siphoned before the pistons were retracted.

 D. The brake calipers were not bled properly.

Answer A is incorrect. A clogged port would not cause the condition described.

Answer B is incorrect. A cracked reservoir would not cause the reservoir to be overfilled.

Answer C is correct. The technician should have siphoned off some fluid before retracting the caliper pistons.

Answer D is incorrect. Bleeding the brakes would lower the fluid, not raise it.

34. Following disassembly of a rear-wheel drum brake disassembly, the drums are found to be badly scored with blue areas on the braking surface. The technician should do all of the following EXCEPT:

 A. inspect the shoes to see if the rivets have contacted and scored the drums.

 B. measure the drums with a Vernier caliper to see if they may be reconditioned on a brake lathe and reused.

 C. inspect the wheel cylinders for brake fluid leakage.

 D. lubricate all self-adjusting hardware with high temperature waterproof grease.

Answer A is incorrect. The drums should be inspected.

Answer B is correct. A micrometer rather than a Vernier caliper should be used to measure the drum's inner diameter to determine if the wear limit would be exceeded.

Answer C is incorrect. The wheel cylinders should be inspected for leaks.

Answer D is incorrect. The self-adjusting hardware contact points should be lubricated.

TASK E.8

35.　Brake drums are to be turned using an off-car brake lathe. Technician A says to keep the lathe speed low when machining shallow cuts on the drums. Technician B says a rubber belt may be used around the circumference of the drum to keep it from getting hot while being machined. Who is correct?

TASK E.9

A.　A only

B.　B only

C.　Both A and B

D.　Neither A nor B

Answer A is correct. Only Technician A is correct. The lathe's rotational speed should not be too fast when machining drums.

Answer B is incorrect. The belt is used to dampen vibration and prevent a loud "ringing" noise while the drum is being machined.

Answer C is incorrect. Only Technician A is correct.

Answer D is incorrect. Only Technician B is incorrect.

36.　The primary (smaller) shoe on a duo-drum brake system should be installed:

TASK E.13

A.　before the secondary shoe.

B.　after the secondary shoe.

C.　toward the front of the vehicle.

D.　toward the rear of the vehicle.

Answer A is incorrect. It makes no difference which shoe is hung first on the backing plate.

Answer B is incorrect. It makes no difference which shoe is hung first on the backing plate.

Answer C is correct. The primary (smaller) shoe goes toward the front of the vehicle.

Answer D is incorrect. The secondary (larger) shoe goes toward the rear of the vehicle

37.　After installation of a vehicle's wheels while on a lift, all of the wheel lug nuts should be:

TASK E.16

A.　tightened by hand before the vehicle is lowered to the ground.

B.　tightened lightly with a wrench before the vehicle is lowered to the ground then fully torqued.

C.　fully tightened with an impact wrench before the vehicle is lowered to the ground.

D.　fully torqued to specs before the vehicle is lowered to the ground.

Answer A is incorrect. The wheels would be too loose when lowered to the ground.

Answer B is correct. The wheels would then be torqued to specs once they are on the ground.

Answer C is incorrect. An air wrench could possibly over-tighten the lug nuts.

Answer D is incorrect. At least some of the wheels would spin freely instead of being held in place while being torqued.

TASK E.19

38. While inspecting the brakes on a disc-brake equipped vehicle, a maintenance technician finds that the brake caliper slides are severely corroded and dry of lubricant. In most cases, the correct action would be to:

    A. disregard their condition.

    B. replace the calipers.

    C. disassemble the brake assemblies; clean and lubricate the slides.

    D. replace the corroded slides with new ones.

    Answer A is incorrect. The condition should be addressed.

    Answer B is incorrect. Replacing the calipers is not normally required for this condition. The slides should be replaced.

    Answer C is incorrect. If corroded, the slides should likely be replaced.

    Answer D is correct. The slides are corroded and would likely need replacing.

TASK E.21

39. A technician finds that a brake caliper has stripped mounting threads. Which of the following should be done?

    A. Use thread restorer on the damaged threads.

    B. Use a tap to restore the threads.

    C. Tear down the caliper and replace the defective part.

    D. Replace the caliper.

    Answer A is incorrect. Using thread restorer would be an unsafe repair procedure.

    Answer B is incorrect. Chasing the threads with a tap would not be a safe repair.

    Answer C is incorrect. Caliper rebuilding is an involved procedure best left to experienced technicians.

    Answer D is correct. The vehicle should be taken out of service until the caliper has been rebuilt, or more likely is replaced.

TASK E.24

40. Rotors are being turned using an on-car brake lathe. Technician A says that using an on-car lathe is quicker than using an off-car brake lathe. Technician B says that some on-car brake lathes use the vehicle's engine to rotate the brake rotor while it is being turned on the vehicle. Who is correct?

    A. A only

    B. B only

    C. Both A and B

    D. Neither A nor B

    Answer A is incorrect. Technician B is also correct.

    Answer B is incorrect. Technician A is also correct.

    Answer C is correct. Both Technicians are correct. An on-car lathe is faster and may use the engine and drive train to rotate the rotors.

    Answer D is incorrect. Neither Technician is incorrect.

41. The purpose of burnishing brake pads is to:

    A. seat the pads in the rotor grooves.
    B. transfer pad material to the rotors.
    C. make certain the brakes are not spongy.
    D. eliminate air from the brake hydraulic system.

TASK E.29

Answer A is incorrect. The purpose of burnishing is not to seat new pads into previously used rotors.

Answer B is correct. Burnishing conditions the rotors for more effective braking.

Answer C is incorrect. Bleeding the brakes would likely fix this condition.

Answer D is incorrect. This is not a true statement; bleeding the brakes does this.

42. A vacuum booster fails to fully assist the power brakes on a vehicle. Technician A says that the vacuum source to the booster should be checked with a vacuum gauge. Technician B says that the vacuum check valve should be checked for leakage. Who is correct?

    A. A only
    B. B only
    C. Both A and B
    D. Neither A nor B

TASK E.31

Answer A is correct. Only Technician A is correct. A vacuum gauge should be used to check the vacuum source.

Answer B is incorrect. The check valve should be inspected and checked for possible leakage or failure if there is no reserve assist, the check valve does not affect normal braking assist.

Answer C is incorrect. Only Technician A is correct.

Answer D is incorrect. Only Technician A is correct.

43. Technician A says that with some electro-hydraulic braking systems, the high-pressure reservoir supplies the required brake pressure quickly and precisely to the wheel brakes without driver involvement. Technician B says that some electro-hydraulic braking systems offer improved active safety when braking in a corner or on a slippery surface. Who is correct?

TASK E.33

    A. A only
    B. B only
    C. Both A and B
    D. Neither A nor B

Answer A is incorrect. Technician B is also correct.

Answer B is incorrect. Technician A is also correct.

Answer C is correct. Both Technicians are correct.

Answer D is incorrect. Neither Technician is incorrect.

TASK F.2

44. While checking for voltage drop in a charging system while the charging system is operating, a technician finds 0.9 volts being dropped between the alternator and the battery on the insulated side. Which of the following is the most likely cause?

   A.   A faulty voltage regulator is causing the voltage drop.
   B.   The alternator's B+ ring terminal connection has burned where the cable is attached.
   C.   The battery ground cable is corroded.
   D.   This is an acceptable amount of voltage drop.

   Answer A is incorrect. A regulator does not cause voltage drop in a circuit.

   Answer B is correct. The burnt condition of the connector offers resistance to current flow and therefore drops voltage.

   Answer C is incorrect. The battery ground cable does not go to the alternator.

   Answer D is incorrect. 0.9 volts is excessive; no more than 0.2 volts at full charge would be acceptable.

TASK F.4

45. Which of the following could cause unwanted resistance in a tail light circuit?

   A.   A larger-than-specification tail lamp
   B.   The addition of a trailer harness
   C.   The addition of an additional stop/tail light on a trailer hitch receiver
   D.   Water leakage into a tail light housing

   Answer A is incorrect. A larger lamp would offer lower resistance.

   Answer B is incorrect. A harness added would lower total taillight circuit resistance.

   Answer C is incorrect. The extra stop/tail light would lower total taillight circuit resistance

   Answer D is correct. Water would cause rust and corrosion, leading to high resistance at the tail light socket.

TASK F.7

46. A standard flooded lead-acid battery is being replaced with an absorbed glass mat (AGM) battery. Technician A says that an AGM battery is less rugged than a flooded cell battery. Technician B says that under normal conditions, an AGM battery does not emit harmful hydrogen gas. Who is correct?

   A.   A only
   B.   B only
   C.   Both A and B
   D.   Neither A nor B

   Answer A is incorrect. Only Technician B is correct.

   Answer B is correct. Only Technician B is correct. An AGM battery is in fact more robust and does not gas under normal circumstances.

   Answer C is incorrect. Only Technician B is correct.

   Answer D is incorrect. Only Technician A is incorrect.

47. Which of the following is best suited for fully charging a completely discharged 12-volt battery?

TASK F.8

A. Charge the battery for at least 30 minutes by racing the vehicle's engine.
B. Charge the battery for at least 1 hour by idling the engine.
C. Remove the battery from the vehicle and trickle charge it for 6 hours.
D. Remove the battery from the vehicle and slow charge it overnight.

Answer A is incorrect. A completely discharged battery should be removed from the vehicle and charged.

Answer B is incorrect. Charging a dead battery in the vehicle can overtax the alternator diodes and eventually cause an alternator failure.

Answer C is incorrect. A 6-hour trickle charge may not be adequate.

Answer D is correct. A good battery will accept an overnight slow charge from a correctly programmed battery charger, and prevent stressing of the vehicle's alternator.

48. Which of the following is LEAST LIKELY to cause excessive starter current draw?

TASK F.11

A. A partially seized engine
B. A shorted field winding
C. Faulty armature brushes
D. A shorted armature

Answer A is incorrect. A partially seized engine would cause current draw to be high.

Answer B is incorrect. A shorted field would cause current draw to be high.

Answer C is correct. Worn brushes would offer greater resistance and lower amperage draw through the starter armature.

Answer D is incorrect. A shorted armature would cause current draw to be higher than normal.

49. Before a starter is removed from an engine, it's important to:

TASK F.13

A. spot the engine to cylinder #1 TDC.
B. spot the engine to cylinder #1 BDC.
C. disconnect the positive lead from the battery.
D. disconnect the negative lead from the battery.

Answer A is incorrect. Spotting the engine at TDC is not necessary.

Answer B is incorrect. Spotting the engine at BDC is not necessary.

Answer C is incorrect. The positive lead need not be disconnected from the battery.

Answer D is correct. The negative ground lead must be disconnected from the battery.

TASK F.18

50. The right side headlight fails a safety inspection because it does not work. Which of the following is the LEAST LIKELY cause?

   A. The bulb is burned out.

   B. The socket is corroded.

   C. The right side fuse has blown.

   D. The bulb is the incorrect type.

   Answer A is incorrect. The bulb may have burned out.

   Answer B is incorrect. The socket may have corroded if water got into the headlight bucket.

   Answer C is incorrect. The fuse may have blown if water got into the headlight socket and caused a short.

   Answer D is correct. It is unlikely that an incorrect headlight would be installed, and remain installed, if it did not work.

TASK F.21

51. A horn fails to operate as it should. Technician A says the horn relay may be stuck closed. Technician B says the horn button may be stuck closed. Who is correct?

   A. A only

   B. B only

   C. Both A and B

   D. Neither A nor B

   Answer A is incorrect. The horn would stay ON

   Answer B is incorrect. The horn would stay ON.

   Answer C is incorrect. Neither Technician is correct.

   Answer D is correct. Neither Technician is correct. Either condition would cause the horn to stay ON.

TASK G.1

52. An A/C clutch fails to operate when the A/C is turned to ON. Any of the following could be the cause EXCEPT:

   A. the A/C system is low on refrigerant.

   B. the A/C clutch fuse has blown.

   C. the A/C clutch has a shorted winding.

   D. the A/C cooling fan fails to operate.

   Answer A is incorrect. Low refrigerant would cause the system to not operate.

   Answer B is incorrect. A blown A/C clutch fuse could be the problem.

   Answer C is incorrect. A shorted clutch winding could be the cause of the blown fuse (described above) .

   Answer D is correct. The fan should operate, but even if it didn't, it would not prevent the A/C compressor clutch from engaging.

53. Which of the following is supposed to cool and return refrigerant from a gaseous state to a liquid state?

    A. The condenser
    B. The A/C compressor
    C. The evaporator
    D. The receiver-drier

TASK G.3

Answer A is correct. The condenser cools the refrigerant to a liquid from its gaseous state from the compressor.

Answer B is incorrect. The compressor does not return the refrigerant to a liquid state.

Answer C is incorrect. The evaporator works opposite to the condenser.

Answer D is incorrect. The receiver-drier traps any residual condensed refrigerant and also filters it.

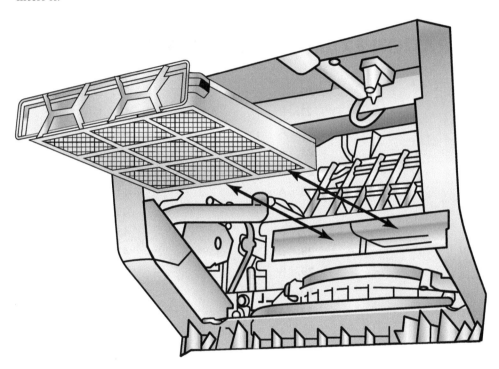

54. Which of the following parts is being replaced in the illustration above?

    A. The cabin air filter
    B. The A/C evaporator
    C. The intake air filter
    D. The glove box

TASK G.4

Answer A is correct. The part shown is a cabin air filter.

Answer B is incorrect. The part is shown is not an evaporator.

Answer C is incorrect. The intake air filter is near the engine, not in the dashboard.

Answer D is incorrect. The glove box is not shown in this illustration.

TASK G.5

55. The A/C serpentine drive belt keeps jumping off its pulley whenever the A/C is turned on. Which of the following is the most likely cause?

A. A pulley or bracket is belt is misaligned.
B. The belt is too small.
C. The belt is made of the wrong material.
D. The belt is too old.

Answer A is correct. Misalignment causes the belt to come off the pulley.

Answer B is incorrect. A too-small belt would not jump off the pulley, even if it fit.

Answer C is incorrect. Incorrect belt material is unlikely to be the cause.

Answer D is incorrect. Even though an old belt may wear, slip, or even break, it is unlikely to routinely jump off its pulley.

PREPARATION EXAM 5—ANSWER KEY

1.	A	20.	C	39.	D
2.	C	21.	B	40.	C
3.	B	22.	A	41.	D
4.	B	23.	A	42.	B
5.	B	24.	B	43.	C
6.	C	25.	A	44.	D
7.	B	26.	D	45.	A
8.	C	27.	B	46.	C
9.	B	28.	D	47.	D
10.	C	29.	B	48.	C
11.	B	30.	C	49.	D
12.	A	31.	A	50.	B
13.	C	32.	A	51.	C
14.	C	33.	D	52.	B
15.	D	34.	C	53.	D
16.	B	35.	D	54.	C
17.	B	36.	C	55.	A
18.	C	37.	D		
19.	B	38.	D		

PREPARATION EXAM 5—EXPLANATIONS

1. Before starting work on any vehicle, it's important to:

    A. gather as much information as you can.

    B. check with the customer.

    C. check with the shop supervisor.

    D. raise the vehicle on the lift.

TASK A.1

Answer A is correct. Without information, the technician would be without clues as to how to proceed.

Answer B is incorrect. The customer has already been interviewed by the service writer.

Answer C is incorrect. The repair order and related vehicle service information should be consulted.

Answer D is incorrect. There may be no reason to raise the vehicle on the lift just yet.

2. While listening to a manual transmission vehicle engine as it is idling, the technician notes a loud ticking noise coming from the upper part of the engine. Which of the following is the most likely cause?

    A. A rod knock

    B. A worn thrust bearing

    C. A collapsed lifter

    D. A worn main bearing

TASK A.3

Answer A is incorrect. A rod bearing would make a deeper sound from the bottom end of the engine.

Answer B is incorrect. Thrust bearing clearance would contribute to a lower end "thunk" noise when the clutch is depressed.

Answer C is correct. A collapsed lifter would cause excessive valve clearance and make such a noise.

Answer D is incorrect. A worn main bearing would make a deeper sound from the lower end of the engine.

TASK A.6

3. All of the following cooling system components should be regularly inspected and tested EXCEPT for the:

A. radiator.

B. condenser.

C. heater core.

D. pressure cap.

Answer A is incorrect. The radiator should be routinely pressure tested.

Answer B is correct. The condenser is not part of the cooling system.

Answer C is incorrect. The heater core would be pressure tested along with the radiator and pressure cap.

Answer D is incorrect. The radiator cap should be routinely pressure tested.

TASK A.8

4. What is the purpose of the tool illustrated above?

A. To remove worm-type hose clamps

B. To remove flat-band type hose clamps

C. To remove wire-type hose clamps

D. To remove double-wire-type hose clamps

Answer A is incorrect. This tool does not remove screw-type hose clamps.

Answer B is correct. The tool shown is a pair of flat-band (also known as constant tension) hose clamp pliers.

Answer C is incorrect. A different type of pliers would be used for wire-type hose clamps.

Answer D is incorrect. A different kind of pliers is used for double-wire-type hose clamps.

5. Which of the following is commonly found on vehicles built through the 80s?

   A. Electric water pumps

   B. Belt-driven water pumps

   C. Vacuum-operated windshield wipers

   D. Draft tubes

TASK A.11

   Answer A is incorrect. Electric water pumps are relatively new.

   Answer B is correct. Most 80's model vehicles use belt-driven water pumps.

   Answer C is incorrect. Vacuum-operated wipers were last seen in 1950s model vehicles.

   Answer D is incorrect. Draft tubes were discontinued with the onset of crankcase emission controls.

6. Technician A says that after many miles, oil can accumulate in the induction system. Technician B says that carbon buildup in the throttle body can make the throttle blade stick closed. Who is correct?

   A. A only

   B. B only

   C. Both A and B

   D. Neither A nor B

TASK A.14

   Answer A is incorrect. Both Technicians are correct.

   Answer B is incorrect. Both Technicians are correct.

   Answer C is correct. Both Technicians are correct. Excessive blow-by can cause oil contamination to build up in the air filter housing. Carbon accumulation in the throttle body can cause the throttle blade to stick closed when the air is moist.

   Answer D is incorrect. Neither Technician is incorrect.

7. Technician A says that one way to check for rusty exhaust pipes is to tap the pipes with ball peen hammer and listen for a metallic "ring." Technician B says that a missing heat shield from a catalytic converter could inadvertently start a fire. Who is correct?

   A. A only

   B. B only

   C. Both A and B

   D. Neither A nor B

TASK A.17

   Answer A is incorrect. Use a light-weight tool such as a screwdriver rather than a hammer.

   Answer B is correct. Only Technician B is correct. Vehicles with missing heat shields when parked in high grass have been known to start grass fires.

   Answer C is incorrect. Only Technician B is correct.

   Answer D is incorrect. Only Technician A is incorrect.

TASK A.20

8. The tool shown in the illustration above is used to disconnect:

    A. brake lines.

    B. heater hoses.

    C. fuel lines.

    D. EVAP hoses.

    Answer A is incorrect. Brake lines use threaded fittings.

    Answer B is incorrect. Heater hoses use clamps.

    Answer C is correct. The tool shown is slid endwise into an in-line fuel connection (or A/C line) to release the internal holding clip.

    Answer D is incorrect. Evaporative emissions hoses are slid on generally without the use of clamps.

TASK A.22

9. Diesel exhaust fluid (DTF) is to be added to a clean diesel vehicle. The cap on the DTF reservoir/bottle is normally:

    A. red.

    B. blue.

    C. white.

    D. yellow.

    Answer A is incorrect. Blue is the correct color of a DTF cap.

    Answer B is correct. Blue is the standard color of the DTF container's cap. Think "AdBlue", the product used by German OEMs.

    Answer C is incorrect. A white cap is not used on a DTF reservoir.

    Answer D is incorrect. A yellow cap is not used for the DTF cap.

10. A test drive reveals that an electronically controlled automatic transmission equipped vehicle starts out in 2nd gear. Technician A says to check for diagnostic trouble codes (DTCs). Technician B says the transmission pan may have to be drained and removed to replace faulty transmission components. Who is correct?

TASK B.1

　　A. A only

　　B. B only

　　C. Both A and B

　　D. Neither A nor B

Answer A is incorrect. Technician B is also correct.

Answer B is incorrect. Technician A is also correct.

Answer C is correct. Both Technicians are correct. The first step after a test drive is to check for DTCs. If DTCs are stored in the TCM, the transmission pan will likely need to be drained and removed in order to service the transmission.

Answer D is incorrect. Both Technicians are correct.

11. Technician A says that automatic transmission fluids (ATFs) are supplied in many different colors. Technician B says that ATF can be smelled to determine if the clutches in the transmission are failing. Who is correct?

TASK B.2

　　A. A only

　　B. B only

　　C. Both A and B

　　D. Neither A nor B

Answer A is incorrect. Red is the standard color for ATF.

Answer B is correct. Only Technician B is correct. You can sniff ATF, checking for a burnt smell to determine if the clutches or bands are slipping, become burnt, and need to be replaced.

Answer C is incorrect. Only Technician B is correct.

Answer D is incorrect. Only Technician A is incorrect.

12. A CV joint boot is found to be dried and nearly split, but not yet leaking lubricant. The boot may be replaced by any of the following methods EXCEPT:

TASK B.4

　　A. replace the boot with a split-type replacement to avoid removing the axle-shaft.

　　B. remove the axle-shaft and install a new boot.

　　C. remove and replace the entire axle-shaft assembly to save time.

　　D. remove, clean and repack the CV joint; install a new boot.

Answer A is correct. Replacing the old boot with a split-type may save labor time and expense but tends to be a temporary repair.

Answer B is incorrect. Removing the axle and replacing the boot may be done, but it takes longer (more labor time).

Answer C is incorrect. Replacing the entire axle-shaft assembly may also be done to save CV joint service labor time.

Answer D is incorrect. Removal, cleaning and regreasing and the CV joint along with replacing the boot may be a viable solution to high cost replacement.

TASK B.5

13. Upon inspection, a leak is found at the transmission cooler line fitting on a vehicle's radiator. Technician A says that tightening the threaded-type fitting to specs may cure the problem. Technician B says that replacing the rubber sealing ring may be all that is needed. Who is correct?

  A.   A only
  B.   B only
  C.   Both A and B
  D.   Neither A nor B

Answer A is incorrect. Technician B is also correct.

Answer B is incorrect. Technician A is also correct.

Answer C is correct. Both Technicians are correct. Various types of fittings are found on transmission cooler lines at the radiator. These include pipe thread brass fittings, fittings with sealing rings, and clip-on fittings.

Answer D is incorrect. Both Technicians are correct.

TASK C.6

14. During a vehicle test drive, a driveline vibration is noticed. Technician A says that the driveshaft may be out of balance. Technician B says the driveshaft may be out of phase. Who is correct?

  A.   A only
  B.   B only
  C.   Both A and B
  D.   Neither A nor B

Answer A is incorrect. Both Technicians are correct.

Answer B is incorrect. Both Technicians are correct.

Answer C is correct. Both Technicians are correct. The driveshaft should be marked before removal to make sure it is re-installed in the same position. It should also be inspected for a loss of the weight normally attached to it.

Answer D is incorrect. Both Technicians are correct.

TASK C.7

15. As shown in the illustration above, driveshaft angles (phasing) are being checked in a vehicle. If they are not both the same angle, which of the following might be likely to happen?

  A.   The clutch might slip.
  B.   The transmission might pop out of gear.
  C.   The analog speedometer needle might fluctuate.
  D.   The driveshaft might vibrate.

Answer A is incorrect. The clutch would not be affected.

Answer B is incorrect. Popping out of gear is not the most likely effect of a driveline being out of phase.

Answer C is incorrect. This is possible, but not the most likely thing to occur.

Answer D is correct. A vibration would likely be felt in the driveline.

16. After a vehicle test drive, differential lubricant is seen leaking at the rear axle flange area and the axle tube is hot. Technician A says that the incorrect differential lubricant may have been used. Technician B says that the axle seal has failed and the axle bearing should be inspected for damage. Who is correct?

TASK C.10

   A. A only
   B. B only
   C. Both A and B
   D. Neither A nor B

   Answer A is incorrect. Incorrect lubricant is not likely to be the cause of this issue.

   Answer B is correct. Only Technician B is correct. The axle seal is obviously leaking and the bearing is heating up due to low lubricant level in the differential/axle tubes.

   Answer C is incorrect. Only Technician B is correct.

   Answer D is incorrect. Only Technician A is incorrect.

17. A broken wheel stud is to be replaced in an axle flange. Which of the following may be required to install the replacement?

TASK C.12

   A. A large ball peen hammer
   B. A press
   C. A pair of slip-joint pliers
   D. A "C" clamp

   Answer A is incorrect. A hammer may be used if there is enough clearance behind the flange, but this is not likely.

   Answer B is correct. If there is not enough room, the axle will have to be removed in order to install a replacement wheel stud. In such cases, a press should be used to properly insert the new wheel stud into an axle flange.

   Answer C is incorrect. Pliers would not be used for this install.

   Answer D is incorrect. A "C" clamp would not be used either.

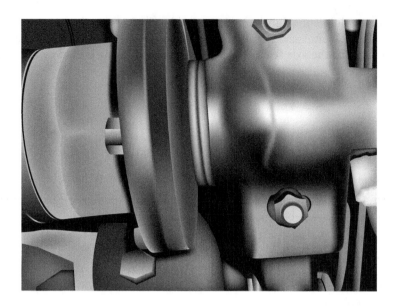

TASK C.18

18. The front end of a 4WD vehicle is being inspected for leaks. Which of the following is likely to be causing the oil residue from a leak, as shown in the illustration above?

    A.   A leaking universal joint

    B.   A leaking gasket

    C.   A leaking seal

    D.   A leaking CV joint

Answer A is incorrect. A universal joint would not leak.

Answer B is incorrect. A transfer case pan gasket would not leak at this location.

Answer C is correct. A leaking seal is the most likely cause.

Answer D is incorrect. A constant velocity (CV) joint is not pictured.

TASK C.19

19. An axle seal is found to be leaking on a 4WD vehicle. Which of the following is the most likely cause?

    A.   Too little transfer case lubricant

    B.   A plugged transaxle vent

    C.   Driving over rough non-road surfaces

    D.   Excessive turning on dry pavement while in 4WD

Answer A is incorrect. Too little lubricant could cause parts failure but not likely a leak.

Answer B is correct. A plugged vent would cause pressure to build up in the transaxle and force lubricant out past the seal.

Answer C is incorrect. The 4WD vehicle is designed for such use.

Answer D is incorrect. This kind of driving is very hard on the 4WD drivetrain, but not on the seals.

20. When a serpentine belt is replaced on a front-wheel drive vehicle, it's important to also make sure all of the following are done EXCEPT:

TASK D.3

   A. make certain that the belt is the correct size and is properly seated in all of the grooved pulleys.

   B. check that the tensioner is free to flex and maintain tension on the belt.

   C. make certain belt dressing is applied to the belt to keep it running quietly and slip-free.

   D. make certain that the tensioner and pulleys do not make any noise as they operate.

   Answer A is incorrect. The belt should seat properly and not be offset to one side.

   Answer B is incorrect. The tensioner should apply spring tension to the belt to keep it firmly in place.

   Answer C is correct. Belt dressing should not be needed when a new belt is installed.

   Answer D is incorrect. All pulleys and tensioner should be silent as the engine idles and the belt rotates.

21. After a power steering pressure hose has been replaced, it's important to:

TASK D.6

   A. bleed the system of fluid.

   B. purge the system of air.

   C. drain and flush the system.

   D. pressurize the system.

   Answer A is incorrect. The power steering system does not require being bled.

   Answer B is correct. It's important to follow the recommended procedures to purge all air out of the PS system after is has been serviced.

   Answer C is incorrect. For a simple hose replacement, flushing the power steering system is not a requirement.

   Answer D is incorrect. There is no need to pressurize the system; the PS pump does this when the engine is started.

22. An idler arm in a pickup truck is worn and needs to be replaced. Which of the following will be needed to complete the job?

TASK D.9

   A. A ball joint separator

   B. A large ball peen hammer

   C. A wheel alignment

   D. New tie rod ends

   Answer A is correct. Either a separator, or a so-called "pickle fork," will be needed to separate the ball joint.

   Answer B is incorrect. A large hammer should not be needed.

   Answer C is incorrect. Simply replacing the idler arm does not necessitate a wheel alignment.

   Answer D is incorrect. New tie rod ends will not be needed for this job.

TASK D.12

23. Ride height is to be checked on a vehicle and compared to specs. Technician A says that ride height is the shortest distance between a level surface and the vehicle. Technician B says that the measurement should be taken with the vehicle loaded with average cargo and a driver. Who is correct?

   A. A only
   B. B only
   C. Both A and B
   D. Neither A nor B

Answer A is correct. Only Technician A is correct. Measure from the level shop floor to the chassis or suspension components as recommended by the manufacturer, taking the lowest and highest measurements and comparing them to specs.

Answer B is incorrect. The car should not be loaded with cargo or passengers.

Answer C is incorrect. Only Technician A is correct.

Answer D is incorrect. Only Technician B is incorrect.

TASK D.14

24. Worn or damaged rebound bumpers often indicate that there is a need to replace the:

   A. springs.
   B. shock absorbers.
   C. upper control arms.
   D. lower control arms.

Answer A is incorrect. Weak springs would allow the vehicle to sag and cause impact against the jounce bumpers.

Answer B is correct. Worn shocks allow the vehicle suspension to rebound in excess and damage the rebound bumpers.

Answer C is incorrect. Unless bent, an upper control arm would not affect wear of rebound bumpers.

Answer D is incorrect. Likewise, a lower control arm would not normally affect the rebound bumpers.

25. The front strut assemblies on a vehicle are being inspected. Technician A says to check that the struts are not bent, broken, or leaking oil. Technician B says when performing the "bounce test" the vehicle should not bounce up and down more than four times. Who is correct?

TASK D.23

   A. A only
   B. B only
   C. Both A and B
   D. Neither A nor B

Answer A is correct. Only Technician A is correct. All of these should be checked. Oil misting is OK, but oil leakage is not OK.

Answer B is incorrect. The bounce test is a legitimate test for older vehicles, but only two bounces are acceptable.

Answer C is incorrect. Only Technician A is correct.

Answer D is incorrect. Only Technician B is incorrect.

26. An older vehicle shows signs of dried and split jounce bumpers. Technician A says that this may cause the vehicle to bounce excessively after hitting a bump. Technician B says that the vehicle may experience unstable tracking. Who is correct?

    TASK D.31

    A. A only

    B. B only

    C. Both A and B

    D. Neither A nor B

    Answer A is incorrect. Jounce bumpers do not affect vehicle bouncing.

    Answer B is incorrect. Jounce bumpers do not affect vehicle tracking.

    Answer C is incorrect. Neither Technician is correct.

    Answer D is correct. Neither Technician is correct. Worn shocks/ struts would affect vehicle bouncing and worn rear suspension parts like ball joints could affect vehicle tracking.

27. With a vehicle's wheels of the ground, the rear struts are found to be loose at their upper end. Which of the following should be done first?

    TASK D.32

    A. Replace the struts.

    B. Inspect the upper mounting plates.

    C. Replace the mounting plates.

    D. Replace the strut bushings.

    Answer A is incorrect. Unless they are worn, leaking, or damaged, the struts do not need to be replaced.

    Answer B is correct. The mounting plates should be inspected for looseness or damage.

    Answer C is incorrect. Check the strut plates before replacing them. Replace both sides as a pair.

    Answer D is incorrect. Strut bushings at the bottom of the strut would not cause this issue.

28. Which of the following would LEAST LIKELY be affected by loose rear tie rods on an independent rear-suspension vehicle?

    TASK D.35

    A. Toe setting

    B. Vehicle tracking

    C. Rear sway

    D. Ride comfort

    Answer A is incorrect. Loose rear tie rods would affect the rear toe setting.

    Answer B is incorrect. Vehicle tracking could be affected by loose rear tie rod ends.

    Answer C is incorrect. Rear sway may be evident due to loose rear tie rods.

    Answer D is correct. Other than possible sway, ride comfort would not likely be affected by loose tie rods.

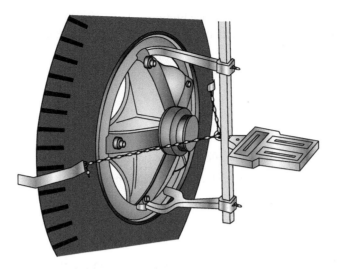

TASK D.**41**

29. Which of the following is being measured using the setup in the illustration above?

A.   Included angle

B.   Camber

C.   Toe

D.   Steering Axis inclination

Answer A is incorrect. Included angle is not being measured.

Answer B is correct. The setup shown is being used to check a wheel's camber measurement.

Answer C is incorrect. Toe is not being measured with this setup.

Answer D is incorrect. SAI is not being checked with this setup.

TASK D.**43**

30. Technician A says most front-wheel drive vehicles use a degree of toe out to enable the front tires to run parallel to each other at road speeds. Technician B says that toe is one of the most critical alignment settings relative to tire wear. Who is correct?

A.   A only

B.   B only

C.   Both A and B

D.   Neither A nor B

Answer A is incorrect. Technician B is also correct.

Answer B is incorrect. Technician A is also correct.

Answer C is correct. Both Technicians are correct. Toe greatly affects tire wear if the tires are not running parallel to each other when the vehicle is traveling straight.

Answer D is incorrect. Neither Technician is incorrect.

31. When a tire is being mounted on a wheel rim, it's important to match mount the tire on the rim:

    A.   to help minimize the final combination of force variation and/or imbalance.
    B.   to avoid damage to the tire pressure monitoring system (TPMS) sending unit.
    C.   so that the tire does not have to be rebalanced.
    D.   to avoid damage to the tire.

    TASK D.53

    Answer A is correct. So-called "match mounting" allows minimizing an "unfavorable stackup" of tire and wheel imbalances.

    Answer B is incorrect. Match mounting has nothing to do with the TPMS sensor.

    Answer C is incorrect. Tires should always be (re)balanced after being mounted on a wheel rim.

    Answer D is incorrect. Damage to the tire is not a concern related to match mounting.

32. A replacement tire pressure monitoring system (TPMS) sensor has been installed in a vehicle's left-rear tire and rim assembly. Technician A says to use a TPMS activator to perform a relearn procedure on the new sensor. Technician B says that the three other tires and the spare will also need a relearn procedure performed. Who is correct?

    TASK D.55

    A.   A only
    B.   B only
    C.   Both A and B
    D.   Neither A nor B

    Answer A is correct. Only Technician A is correct. Only the replaced sensor needs to be undergo a relearn procedure. The other sensors should remain programmed correctly.

    Answer B is incorrect. Only Technician A is correct.

    Answer C is incorrect. Only Technician A is correct.

    Answer D is incorrect. Only Technician B is incorrect.

33. A metal brake line at the rear of a vehicle is found to be dented by a floor jack. Which of the following should be done?

    A.   Replace the dented section using a stainless steel compression fitting.
    B.   Replace the dented section using a brass compression fitting.
    C.   Replace the dented section using a copper compression fitting.
    D.   Replace the dented brake line from fitting to fitting with a steel brake line.

    TASK E.3

    Answer A is incorrect. Compression fittings are not acceptable.

    Answer B is incorrect. Compression fittings are not acceptable.

    Answer C is incorrect. Compression fittings are not acceptable.

    Answer D is correct. Given the other options, the entire brake line should be replaced.

34. A brake warning light bulb fails to operate. Technician A says the warning brake light bulb may be faulty. Technician B says the park brake light switch may be misadjusted. Who is correct?

    A.   A only
    B.   B only
    C.   Both A and B
    D.   Neither A nor B

    TASK E.5

    Answer A is incorrect. Technician B is also correct.

    Answer B is incorrect. Technician A is also correct.

    Answer C is correct. Both Technicians are correct. Either the bulb could be blown or the switch could be misadjusted.

    Answer D is incorrect. Neither Technician is incorrect.

TASK E.11

35. Which of the following should be used when lubricating brake shoe support pads?

    A. Vaseline®
    B. Chassis lubricant
    C. WD-40®
    D. Water/heat resistant grease

    Answer A is incorrect. Vaseline® is the incorrect lubricant.

    Answer B is incorrect. While it may work, chassis lubricant may not meet the OEM's recommendations.

    Answer C is incorrect. WD-40® is too thin and would likely be washed off.

    Answer D is correct. Use only OEM recognized water and heat resistant lubricant designed for such applications.

TASK E.12

36. Technician A says a wheel cylinder can be rebuilt using a replacement cups and boots kit. Technician B says many wheel cylinder housings are made from aluminum making honing not possible, so the entire wheel cylinder assembly should be replaced. Who is correct?

    A. A only
    B. B only
    C. Both A and B
    D. Neither A nor B

    Answer A is incorrect. Technician B is also correct.

    Answer B is incorrect. Technician A is also correct.

    Answer C is correct. Both Technicians are correct. The wheel cylinder cannot be honed if aluminum and rebuild kits can be difficult to find for some wheel cylinders. The other most cost effective option is to replace the entire wheel cylinder, which is commonly done.

    Answer D is incorrect. Neither Technician is incorrect.

TASK E.15

37. Parking brake cables should be lubricated with:

    A. Permatex®.
    B. WD-40®.
    C. Wheel bearing grease.
    D. Graphite lubricant.

    Answer A is incorrect. Permatex® is a sealant, not a lubricant.

    Answer B is incorrect. WD-40® is too thin.

    Answer C is incorrect. Wheel bearing grease would harden in cold temperatures.

    Answer D is correct. Graphite lubricant would be a good choice.

38. A disc brake caliper is being dismounted for brake pad replacement. Technician A says the caliper may be hung by the brake hose during the repair. Technician B says that an air gun may be used to remove brake dust from the caliper. Who is correct?

TASK E.18

   A. A only
   B. B only
   C. Both A and B
   D. Neither A nor B

Answer A is incorrect. Technician B is also incorrect.

Answer B is incorrect. Technician A is also incorrect.

Answer C is incorrect. Neither Technician is correct.

Answer D is correct. Neither Technician is correct. The caliper should never be allowed to hang from the brake line; it should be suspended from a stiff wire hooked to a suspension component. Shop air should never be used to remove brake dust; doing so would spread dangerous brake dust to the surrounding air for possible inhalation.

39. To remove a hubless brake rotor from a vehicle with disc brakes, the technician should first:

TASK E.23

   A. unbolt the lug nuts, remove the wheel assembly and lift the rotor off the wheel studs.
   B. unbolt the lug nuts, remove the wheel assembly and remove the castellated nut and slide the rotor off the spindle with the outer bearing parts.
   C. unbolt the lug nuts, remove the wheel assembly and use a slide hammer to remove the hub assembly.
   D. unbolt the lug nuts, remove the wheel assembly and unbolt and swing the caliper aside.

Answer A is incorrect. The rotor must first be removed.

Answer B is incorrect. The hubless rotor does not include bearing assemblies.

Answer C is incorrect. The hub stays in place.

Answer D is correct. The brake caliper must first be removed in order to get a brake rotor off the vehicle.

40. A parking brake using an integral-style caliper is to be adjusted. Technician A says to depress the service brake pedal several times. Technician B says to check and, if necessary, to adjust the parking brake cable to the proper length. Who is correct?

TASK E.26

   A. A only
   B. B only
   C. Both A and B
   D. Neither A nor B

Answer A is incorrect. Technician B is also correct

Answer B is incorrect. Technician A is also correct

Answer C is correct. Both Technicians are correct. Depressing the brake pedal sets the pads into position. The cable length should be checked and adjusted if necessary.

Answer D is incorrect. Neither Technician is incorrect.

TASK E.28

41. Wheel lug nuts should be properly torqued to specs for all of the following reasons EXCEPT:

   A.   to help center the wheel.

   B.   to make sure the wheel does not come off.

   C.   to prevent brake rotor warpage.

   D.   To avoid wheel imbalance.

   Answer A is incorrect. Tightening and torquing the wheel helps to center it.

   Answer B is incorrect. The lugs nuts would not loosen and come off if properly torqued.

   Answer C is incorrect. The brake rotors could become warped if the lug nuts are not properly torqued.

   Answer D is correct. Uneven torque of the lug nuts would not cause a wheel to be out of balance.

TASK E.30

42. To test for brake pedal travel, push the brake pedal a few times to relieve vacuum in the brake booster, hold the brake pedal down firmly, and:

   A.   turn on the ignition and note the change of brake pedal position.

   B.   start the engine and note the amount of change to the brake pedal position.

   C.   start and rev up the engine and note the change of brake pedal position.

   D.   accelerate and hold the engine at 1500 rpm and note the brake position change.

   Answer A is incorrect. This does nothing to check the booster or freeplay.

   Answer B is correct. If the amount of freeplay and the booster operation are correct, the pedal should drop a bit once the engine is started.

   Answer C is incorrect. Revving the engine is not necessary.

   Answer D is incorrect. Accelerating the engine is not required.

TASK E.32

43. With a vehicle idling, a loud hiss is heard and the engine stalls whenever the brake pedal is applied. Which of the following is the most likely cause?

   A.   A faulty brake master cylinder

   B.   A leaking vacuum line to the brake booster check valve

   C.   A leaking brake booster

   D.   A misadjusted brake pedal pushrod

   Answer A is incorrect. A master cylinder would not affect an engine idle.

   Answer B is incorrect. The vacuum line upstream from the check valve would not cause the problem

   Answer C is correct. If a brake booster is leaking, engine vacuum is lost and cause the engine to stumble or stall.

   Answer D is incorrect. An incorrectly adjusted pushrod would cause dragging brakes or other symptoms.

44. When the ignition system of a vehicle is first turned on, the supplemental restraint system (SRS/airbag) light comes on and then goes off. This indicates that the SRS:

TASK F.1

    A.   lamp burned out.

    B.   fuse is blown.

    C.   has a fault and has stored a DTC.

    D.   self-test has been successfully completed.

Answer A is incorrect. The lamp did not burn out.

Answer B is incorrect. The fuse did not blow.

Answer C is incorrect. This does not indicate a fault has been stored.

Answer D is correct. This is how the self test is supposed to perform.

45. Which of the following would be LEAST LIKELY to cause keep alive memory (KAM) of engine parameters to be lost?

TASK F.6

    A.   A blown fuse for the EBCM (ABS) module

    B.   A burned fusible link

    C.   A blown fuse for the powertrain control module (PCM)

    D.   A dead battery

Answer A is correct. A blown EBCM fuse would not likely affect the engine circuit and a loss of engine KAM.

Answer B is incorrect. This would most likely cause engine KAM to be erased.

Answer C is incorrect. This would likely cause engine KAM to be erased

Answer D is incorrect. A dead battery would likely cause most, if not all, KAM to be erased.

46. The B+ battery cable connection at the battery appears to be clean, yet a voltage drop test reveals a 1.6 volt drop in the B+ cable to the starter when the engine is being cranked. Technician A says the B+ cable may be corroded under its insulation. Technician B says the B+ connection at the starter may be oil soaked or loose. Who is correct?

TASK F.9

    A.   A only

    B.   B only

    C.   Both A and B

    D.   Neither A nor B

Answer A is incorrect. Technician B is also correct.

Answer B is incorrect. Technician A is also correct.

Answer C is correct. Both Technicians are correct. Either of these conditions could cause the voltage drop as measured.

Answer D is incorrect. Neither Technician is incorrect.

47. Which of the following best describes the last connection to be made while jump starting a vehicle with a dead battery using another vehicle:

   A. Connect the positive to the positive battery terminal of vehicle to be jump started.
   B. Connect the positive to the negative battery terminal of a vehicle to be jump started.
   C. Connect the positive to the engine block of the vehicle to be jump started.
   D. Connect the negative to the engine block of the vehicle to be jump started.

   Answer A is incorrect. This is not the last step in the process.

   Answer B is incorrect. This is never a step when jump starting a vehicle with another vehicle.

   Answer C is incorrect. This is never a step when jump starting a vehicle with another vehicle.

   Answer D is correct. Connecting the negative to the engine block for a ground connection is the last step when jumping a vehicle with another vehicle.

48. A neutral safety switch is suspected of being faulty. Technician A says that an easy way to bypass the suspected switch would be to use a jumper wire across the starter relay terminals at the fuse panel. Technician B says the entire starter control circuit can be bypassed by using a remote starter switch at the starter solenoid. Who is correct?

   A. A only
   B. B only
   C. Both A and B
   D. Neither A nor B

   Answer A is incorrect. Technician B is also correct

   Answer B is incorrect. Technician A is also correct.

   Answer C is correct. Both Technicians are correct. The neutral safety switch can be bypassed at the relay box, and the entire control circuit can be bypassed by "hot wiring" the starter solenoid.

   Answer D is incorrect. Both Technicians are correct.

49. A vehicle's alternator is to be removed from a vehicle for testing. The first step is to remove the:

   A. two or three mounting bolts holding the alternator to the bracket.
   B. alternator bracket.
   C. battery positive cable connector from the alternator.
   D. battery negative cable connector at the battery.

   Answer A is incorrect. Power from the battery should be disconnected first.

   Answer B is incorrect. The bracket should not need to be removed.

   Answer C is incorrect. The alternator B+ connector is live, so removing it first could result in a short to ground.

   Answer D is correct. Before doing anything else, it's important to first remove the battery's negative cable connector from the battery.

50. A vehicle's headlight lens is fogged over with moisture on the inside. Which of the following is the LEAST LIKELY cause?

TASK F. 19

 A. A small stone-impact hole in the lens
 B. A crack at the base of the headlight bulb
 C. A missing "O" ring at the headlight socket
 D. A faulty gasket around the perimeter of the lens

Answer A is incorrect. This is a likely cause for moisture to get in.

Answer B is correct. A cracked at the bulb's base is the least likely cause for moisture in the headlight assembly.

Answer C is incorrect. A missing "O" ring from the bulb's base could be the cause.

Answer D is incorrect. A faulty lens gasket could be the cause.

51. A front wiper blade's rubber is found to be torn and hanging from the wiper blade. Technician A says that just the rubber insert can be replaced. Technician B says it's faster and easier to replace the entire wiper blade. Who is correct?

TASK F.22

 A. A only
 B. B only
 C. Both A and B
 D. Neither A nor B

Answer A is incorrect. Technician B is also correct.

Answer B is incorrect. Technician A is also correct.

Answer C is correct. Both Technicians are correct. Even if inserts are available, it's easier, quicker, and more reliable to replace the entire wiper blade.

Answer D is incorrect. Both Technicians are correct.

52. Whenever a vehicle is heavily accelerated, air blowing from the dashboard vents shifts to the defroster vents. Which of the following is the most likely cause?

TASK G.1

 A. A defective blower motor
 B. A faulty vacuum check valve
 C. A faulty blower switch
 D. A faulty blend door sensor

Answer A is incorrect. The blower works fine and is not at fault.

Answer B is correct. If the air door's position changes when engine vacuum is lost (such as during hard acceleration), the HVAC system's vacuum check valve is likely at fault.

Answer C is incorrect. The blower speed did not change, so the switch is fine.

Answer D is incorrect. The air temperature did not change, just the position of the door.

TASK G.2

53. A leak is detected at the lower left corner of an A/C condenser. Technician A says the leak can be fixed by soldering it. Technician B says the leak can be fixed by using an epoxy sealant. Who is correct?

A. A only

B. B only

C. Both A and B

D. Neither A nor B

Answer A is incorrect. Technician B is also incorrect.

Answer B is incorrect. Technician A is also incorrect.

Answer C is incorrect. Neither Technician is correct.

Answer D is correct. Neither Technician is correct. The condenser cannot be reliably repaired and should be replaced.

TASK G.4

54. If a cabin air filter is to be replaced, it may be found located either under the dashboard or:

A. under the vehicle.

B. under the passenger's seat.

C. inside the engine compartment.

D. in the headliner above the passenger's seat.

Answer A is incorrect. A cabin filter would not be located under the vehicle.

Answer B is incorrect. This is not where you would find the cabin air filter.

Answer C is correct. Some vehicles have the cabin filter located in the engine compartment or even outside the vehicle by the wiper linkage.

Answer D is incorrect. This is not where you would find the cabin filter.

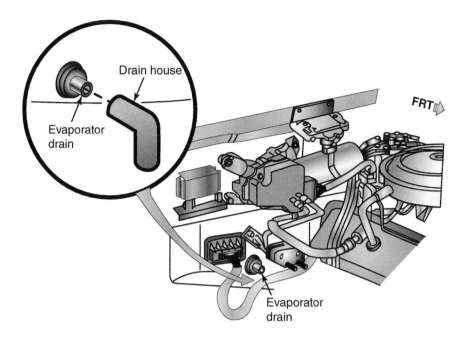

55. The drain hose illustrated above, drips water when the A/C is ON, but not at other times. Which of the following is the most likely cause?

TASK G.6

   A. Condensate is draining from the evaporator.

   B. Water is leaking from the evaporator.

   C. There is a leak in the heater core.

   D. The condenser is leaking.

   Answer A is correct. It is normal for condensate from the evaporator to drip from a drain hose when the A/C is operating.

   Answer B is incorrect. The evaporator is not leaking.

   Answer C is incorrect. The heater core is not leaking.

   Answer D is incorrect. The condenser is located elsewhere and does not drip condensate.

PREPARATION EXAM 6—ANSWER KEY

1.	D	20.	D	39.	B
2.	B	21.	B	40.	C
3.	C	22.	C	41.	C
4.	C	23.	B	42.	C
5.	C	24.	B	43.	C
6.	A	25.	C	44.	D
7.	C	26.	D	45.	C
8.	B	27.	A	46.	D
9.	D	28.	A	47.	D
10.	C	29.	A	48.	D
11.	B	30.	C	49.	D
12.	A	31.	C	50.	C
13.	B	32.	D	51.	B
14.	A	33.	A	52.	D
15.	D	34.	A	53.	A
16.	C	35.	B	54.	A
17.	C	36.	B	55.	B
18.	A	37.	C		
19.	C	38.	A		

PREPARATION EXAM 6—EXPLANATIONS

TASK A.1

1. A vehicle's exhaust emits blue smoke. Which of the following is the most likely cause?

   A. A dirty air filter
   B. A stuck-open crankcase ventilation system
   C. A faulty spark plug
   D. Leaking valve stem seals

   Answer A is incorrect. A dirty air filter could choke off incoming air, but not cause blue smoke.

   Answer B is incorrect. A stuck-open PCV valve might cause a rough engine idle, but not blue smoke.

   Answer C is incorrect. A faulty spark plug would not contribute to blue smoke.

   Answer D is correct. Oil is leaking down the valve stem and into the combustion chamber.

2. Gasoline is leaking from an older vehicle's metal fuel tank. Technician A says the fuel tank may have been over-filled. Technician B says the fuel tank may have rusted from water lying at the bottom of the tank. Who is correct?

TASK A.2

   A. A only

   B. B only

   C. Both A and B

   D. Neither A nor B

Answer A is incorrect. Within limits, over-filling would only inundate the EVAP canister.

Answer B is correct. Only Technician B is correct. Water in the metal tank may have rusted it out to the point of leaking.

Answer C is incorrect. Only Technician B is correct.

Answer D is incorrect. Only Technician A is incorrect.

3. Technician A says that some coolant recovery tanks should not be opened until cooled off. Technician B says that a pressure test would confirm if a radiator cap cannot hold pressure. Who is correct?

TASK A.6

   A. A only

   B. B only

   C. Both A and B

   D. Neither A nor B

Answer A is incorrect. Technician B is also correct.

Answer B is incorrect. Technician A is also correct.

Answer C is correct. Both Technicians are correct. Some coolant recovery bottles are not to be opened when the coolant is hot because of the possibility of being scalded. A pressure test would confirm if the radiator cap leaks pressure.

Answer D is incorrect. Neither Technician is incorrect.

4. A cooling system thermostat is being replaced. Which of the following should be done when installing a new one?

TASK A.9

   A. The gasket should be liberally coated with RTV sealant.

   B. The thermostat housing should also be replaced.

   C. The small air bleed hole should be located at the top.

   D. The thermostat should be adjusted to open at 180 degrees Fahrenheit.

Answer A is incorrect. Only a dab of RTV should be spread to hold the gasket in place during installation of the thermostat housing.

Answer B is incorrect. There is no need to replace the thermostat housing unless it is faulty.

Answer C is correct. The small air bleed hole should be positioned at the top in order for air to automatically purge as coolant circulates.

Answer D is incorrect. The thermostat is factory set and cannot be adjusted.

TASK A.11

5. A timing belt with 100,000 miles on it is being replaced. Technician A says to replace the water pump at the same time. Technician B says to replace the tensioner at the same time. Who is correct?

    A.   A only

    B.   B only

    C.   Both A and B

    D.   Neither A nor B

Answer A is incorrect. Technician B is also correct.

Answer B is incorrect. Technician A is also correct.

Answer C is correct. Both Technicians are correct. The possibly well-worn water pump should be replaced at the same time as the belt. Otherwise the belt would have to be removed again to replace the old water pump when it fails later. The same logic applies to the tensioner (and/or the pulley).

Answer D is incorrect. Neither Technician is incorrect.

TASK A.14

6. Whenever the weather is rainy, an engine's throttle blade sticks closed, making it difficult to accelerate off idle. Which of the following is the most likely cause?

    A.   Carbon is deposited on the throttle body throat and the throttle blade.

    B.   The throttle cable is sticking.

    C.   The throttle cable pulley is warped.

    D.   The cruise control cable is binding the throttle cable.

Answer A is correct. The wet weather is causing the carbon deposits to become sticky and hold the throttle butterfly closed.

Answer B is incorrect. The throttle cable could stick at any time.

Answer C is incorrect. A warped cable pulley is very unlikely.

Answer D is incorrect. The two cables binding together is very unlikely.

TASK A.16

7. A PCV valve does not rattle when it is shaken. Which of the following would be best to do?

    A.   Soak it in a parts cleaning tank.

    B.   Spray it with carburetor or fuel injector cleaner.

    C.   Replace it.

    D.   Use it as it is; there is nothing wrong with the valve.

Answer A is incorrect. Soaking it may not clean it for long. Simply replace it.

Answer B is incorrect. Spraying it may not completely clean it. Again, replace it.

Answer C is correct. Replace it. The labor time spent to clean it is wasteful.

Answer D is incorrect. The PCV valve should rattle.

8. Upon removal and inspection of an engine's spark plugs, oily deposits are found on the insulator of one spark plug. Which of the following is the most likely cause?

TASK A.19

   A. The spark plug is faulty.

   B. A valve stem seal is leaking.

   C. The piston rings are worn.

   D. A spark plug coil wire is faulty.

Answer A is incorrect. The spark plug itself is not at fault.

Answer B is correct. Only one valve stem seal is leaking, thus only one spark plug is fouled.

Answer C is incorrect. Were this the case, all of the plugs would be oil fouled.

Answer D is incorrect. A faulty coil wire might cause misfire related to carbon buildup, but would not cause oil fouling.

9. Which of the following names is commonly used when discussing diesel exhaust fluid (DEF)?

TASK A.22

   A. Adware

   B. Amsol

   C. AdBlut

   D. AdBlue

Answer A is incorrect. Adware is not a brand name for DEF.

Answer B is incorrect. Amsol is not a brand name for DEF.

Answer C is incorrect. ADBlut is not a brand name for DEF.

Answer D is correct. DEF is commonly called AdBlue because it is a commonly sold brand of DEF.

10. During a test drive, an older model automatic transmission vehicle is found to not shift properly. Any of the following could be the cause EXCEPT:

TASK B.1

   A. low transmission fluid.

   B. too much transmission fluid.

   C. an inoperative torque converter clutch.

   D. worn plastic teeth on a governor drive gear.

Answer A is incorrect. Low fluid would influence proper shifting.

Answer B is incorrect. Too much fluid could foam up and affect proper shifting.

Answer C is correct. The torque converter clutch (TCC) would not affect transmission shifting.

Answer D is incorrect. Worn plastic teeth on a mechanical governor would delay up-shifting.

TASK B.3

11. Which of the following would cause an automatic transmission to leak fluid from the bell housing?

   A.   A faulty pan gasket
   B.   A faulty front pump seal
   C.   A faulty rear seal
   D.   A leaking seal at the shift linkage

   Answer A is incorrect. A leaking pan gasket would drip elsewhere further back from the bell housing.

   Answer B is correct. A leak at the bell housing could likely be caused by a leaking front pump seal.

   Answer C is incorrect. A rear seal would leak at the transmission's tail shaft.

   Answer D is incorrect. A leaking seal at the shift linkage would likely drip towards one side of the transmission.

TASK B.5

12. Technician A says a transmission cooler may be located under the car. Technician B says an additional transmission cooler may be installed behind the radiator. Who is correct?

   A.   A only
   B.   B only
   C.   Both A and B
   D.   Neither A nor B

   Answer A is correct. Only Technician A is correct. Some OEMs place the transmission cooler under the vehicle near the transmission.

   Answer B is incorrect. An auxiliary transmission cooler would be installed in front of the radiator, not behind it.

   Answer C is incorrect. Only Technician A is correct

   Answer D is incorrect. Only Technician B is incorrect.

TASK B.6

13. Worn engine mounts are to be replaced. All of the following steps may need to be done EXCEPT:

   A.   raise the engine with an engine hoist.
   B.   disconnect the upper ball joints.
   C.   remove the lower cross-member.
   D.   drop the power steering rack from its mountings.

   Answer A is incorrect. Lifting and supporting the engine is a necessary step.

   Answer B is correct. Upper ball joint removal or disconnect is not necessary when replacing engine mounts even on vehicles with an engine cradle.

   Answer C is incorrect. The lower cross member may need to be removed to gain access to the engine mounts.

   Answer D is incorrect. The rack may need to be lowered in order to gain access to the engine mounts.

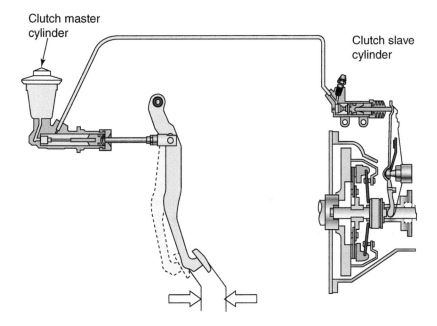

14. At which of the locations illustrated above is air to be bled from the hydraulic clutch system?

   A. From the clutch slave cylinder
   B. From the fluid reservoir
   C. From the clutch master cylinder
   D. From the master cylinder

TASK C.1

   Answer A is correct. In this illustration the bleed screw is on the clutch slave cylinder.

   Answer B is incorrect. Air does not need to be bled from the reservoir.

   Answer C is incorrect. Air does not normally need to be bled from the master cylinder unless it has been replaced. In such a case it would first be "bench bled."

   Answer D is incorrect. Bleeding the master cylinder would not bleed air from the clutch hydraulic system.

15. All of the following are used as either manual transmission or transaxle gear lubricants EXCEPT:

   A. 30 weight motor oil.
   B. 80W90 weight gear oil.
   C. Hypoid gear oil.
   D. P.S. fluid.

TASK C.5

   Answer A is incorrect. 30 weight oil is used by Honda® in some of their manual transmissions.

   Answer B is incorrect. 80W90 weight gear oil is commonly used in manual transmissions.

   Answer C is incorrect. Most lubricants for manual gearboxes (and differentials) are hypoid-type gear oils which contain extreme pressure (EP) additives and anti-wear additives.

   Answer D is correct. Power steering fluid is not to be used in manual transmissions.

TASK C.9

16. A solid rear-axle is leaking lubricant past its seal onto drum-style wheel brakes. Technician A says the axle will need to be removed to replace the seal. Technician B says the brakes at both rear wheels will need to be replaced. Who is correct?

    A.   A only

    B.   B only

    C.   Both A and B

    D.   Neither A nor B

Answer A is incorrect. Technician B is also correct.

Answer B is incorrect. Technician A is also correct.

Answer C is correct. Both Technicians are correct. The flanged axle needs to be removed in order for the replacement seal to be installed. The oil-soaked shoes need to be replaced. Brakes are replaced as a set at either the front or rear of the vehicle.

Answer D is incorrect. Neither Technician is incorrect.

TASK C.13

17. If the differential vent on a solid rear-axle vehicle becomes clogged, which of the following would likely happen first?

    A.   The axle bearings could become overheated.

    B.   The axle bearings could fail.

    C.   The axle seal(s) could leak differential fluid.

    D.   The axle bearings could make noise.

Answer A is incorrect. A clogged vent would not release pressure build-up, but would not necessarily cause overheating.

Answer B is incorrect. The bearings would not necessarily fail, unless fluid leaks out due to a failed seal.

Answer C is correct. Pressure build-up in the differential could cause the seal(s) to leak oil; then the bearings will overheat for lack of lubricant. As they overheat, the lubricant will be burned off and they will make noise until they ultimately fail.

Answer D is incorrect. The bearings would only make noise and fail if the lubricant is lost due to a failed seal.

TASK C.17

18. The right front CV joint on a 4WD vehicle "clicks" loudly during sharp turns. Technician A says the CV joint needs to be replaced. Technician B says that the right front wheel may ultimately quit steering the vehicle. Who is correct?

    A.   A only

    B.   B only

    C.   Both A and B

    D.   Neither A nor B

Answer A is correct. Only Technician A is correct. The joint is preparing to fail and should be replaced.

Answer B is incorrect. If the joint fails, drive torque to the wheel will be lost, but not the ability to steer.

Answer C is incorrect. Only Technician A is correct.

Answer D is incorrect. Only Technician B is incorrect.

EXAMPLE: P0137 LOW VOLTAGE BANK 1 SENSOR 2

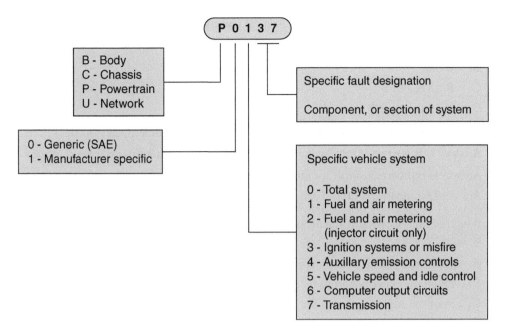

P 0 1 3 7

B - Body
C - Chassis
P - Powertrain
U - Network

0 - Generic (SAE)
1 - Manufacturer specific

Specific fault designation

Component, or section of system

Specific vehicle system

0 - Total system
1 - Fuel and air metering
2 - Fuel and air metering
(injector circuit only)
3 - Ignition systems or misfire
4 - Auxillary emission controls
5 - Vehicle speed and idle control
6 - Computer output circuits
7 - Transmission

19. A failure has occurred in a 4WD vehicle's transmission control module (TCM). Based upon the chart above, which of the following diagnostic trouble codes (DTCs) would be stored?

TASK C.21

A. B0348

B. C0045

C. P0700

D. U0005

Answer A is incorrect. B0348 is a Sun Load Sensor fault code stored in the Body Control Module (BCM).

Answer B is incorrect. C0045 is a LR wheel speed circuit malfunction fault code stored in the BCM.

Answer C is correct. P0700 is a Transmission Control Module (TCM) powertrain-related fault code in the PCM.

Answer D is incorrect. U0005 is a high-speed CAN communication bus failure fault code stored in the BCM.

20. Following replacement of a power steering pump, as well as its pulley, the serpentine drive belt keeps jumping off. Upon investigation, the pulley is noted to be out of alignment by 1/4 inch. Which of the following is the most likely cause?

TASK D.3

A. The pulley is defective.

B. The power steering pump splines are nicked.

C. The power steering bracket is incorrect.

D. The pulley was not pressed all the way onto the PS pump shaft.

Answer A is incorrect. The pulley could be defective, but a more likely cause exists.

Answer B is incorrect. Damaged splines may be the cause, but they would not have allowed the pulley to be pressed onto the PS shaft.

Answer C is incorrect. The bracket was OK when the old power steering pump was replaced.

Answer D is correct. The pulley was not pressed fully onto the PS pump shaft.

TASK D.6

21. Following the replacement of a power steering (PS) pressure hose, the steering system makes a groaning/growling noise when the steering wheel is turned. Technician A says an incorrect PS pressure hose was installed. Technician B says there is air in the PS system. Who is correct?

　　A. A only

　　B. B only

　　C. Both A and B

　　D. Neither A nor B

Answer A is incorrect. The hose in not causing the noise problem; air is in the system.

Answer B is correct. Only Technician B is correct. The system has not been purged of air.

Answer C is incorrect. Only Technician B is correct

Answer D is incorrect. Only Technician A is incorrect.

TASK D.8

22. Which of the following is the proper method of purging air from a power steering (PS) system?

　　A. Prime the power steering system with a pressure bleeder.

　　B. Open the power steering bleed screw and start the engine.

　　C. Start the engine and turn the steering wheel from lock to lock a few times until the noise goes away.

　　D. Drive the vehicle a few miles to get rid of the air in the system.

Answer A is incorrect. A pressure bleeder is used for brakes, not for power steering.

Answer B is incorrect. There is no bleed screw used in the PS system.

Answer C is correct. Turning the steering wheel from lock to lock purges air from the power steering rack.

Answer D is incorrect. Driving the car a few miles is not necessary to purge the air from the PS system.

TASK D.12

23. When the front wheels are turned in either direction on a front McPherson strut equipped vehicle, a squeaking/scrunching noise is heard, and the wheels are hard to turn. Which of the following could be the cause?

　　A. Defective front struts

　　B. Defective upper strut bearings

　　C. Faulty lower ball joints

　　D. Incorrect camber adjustment

Answer A is incorrect. The struts would not make such a noise.

Answer B is correct. The upper strut bearings are dry and need to be replaced.

Answer C is incorrect. The lower ball joints may be faulty, but would not likely make steering difficult.

Answer D is incorrect. Camber may affect how easily the wheels are turned, but an incorrect adjustment would not cause such a noise.

24. All of the following are true statements about replacing a vehicle's upper (non-load carrying) ball joints EXCEPT:

    A. Use a "pickle fork" to loosen the worn ball joints.
    B. Compress the coil springs before replacing the ball joints.
    C. Re-align the front end after replacing the ball joints.
    D. Install a "Zerk" fitting on the ball joint after installation.

    TASK D.16

    Answer A is incorrect. A large fork-like tool known as a pickle fork is hammered into place to separate the ball joint from the lower control arm.

    Answer B is correct. The springs do not need to be compressed to replace non-load carrying ball joints.

    Answer C is incorrect. The wheel alignment is necessary after replacing ball joints.

    Answer D is incorrect. Some ball joints have a "Zerk" type grease fitting on them; The grease fitting is to be installed after the new ball joint is installed.

25. A front steering knuckle is being inspected for damage after a vehicle accident. Technician A says that because of the steering knuckles design, a bent condition may not be easy to see. Technician B says that if a steering knuckle is bent, it might only be apparent when a wheel alignment is attempted. Who is correct?

    A. A only
    B. B only
    C. Both A and B
    D. Neither A nor B

    TASK D.18

    Answer A is incorrect. Technician B is also correct.

    Answer B is incorrect. Technician A is also correct.

    Answer C is correct. Both Technicians are correct. Because of its location and its shape, a bent condition may not be detected until a wheel alignment is attempted.

    Answer D is incorrect. Neither Technician is incorrect.

26. Sway bar bushings are to be replaced. Technician A says that worn sway bar bushings can affect vehicle tracking. Technician B says that following sway bar bushing replacement, a wheel alignment will be needed. Who is correct?

    A. A only
    B. B only
    C. Both A and B
    D. Neither A nor B

    TASK D.22

    Answer A is incorrect. Technician B is also incorrect.

    Answer B is incorrect. Technician A is also incorrect.

    Answer C is incorrect. Neither Technician is correct.

    Answer D is correct. Neither Technician is correct. Sway bar bushings do not affect tracking, nor is an alignment required if they are replaced.

TASK D.25

27. A front wheel bearing is failing due to hard use. Which of the following is the LEAST LIKELY symptom?

   A.   A whining noise will be heard.

   B.   A growling noise will be heard on turns.

   C.   A noise similar to driving on snow tires may be heard.

   D.   The front wheel assembly may experience play when grasped and moved.

   Answer A is correct. A deeper sound is more likely to be heard.

   Answer B is incorrect. A growling sound is typical of a failing wheel bearing.

   Answer C is incorrect. A sound similar to driving on snow tires would be a typical vehicle owner's complaint indicating a possible bad wheel bearing.

   Answer D is incorrect. When the vehicle is raised, grasping the wheel's top and bottom or both sides and shaking it will reveal a loose bearing.

TASK D.28

28. A vehicle with a solid rear axle has a broken panhard rod (tracking bar). Which of the following will the vehicle likely experience?

   A.   Rear end lateral movement or sway

   B.   Poor tire life

   C.   Excessive bouncing

   D.   Drifting at higher speeds

   Answer A is correct. The track bar prevents lateral movement of the vehicle's rear end. The vehicle's back end would tend to sway.

   Answer B is incorrect. Tire life would not be affected.

   Answer C is incorrect. Bouncing would be the fault of faulty shocks or struts.

   Answer D is incorrect. Drifting from side to side at high speed would not likely occur.

TASK D.34

29. The grease fitting on a rear ball joint has receded into the ball joint. Technician A says that the wear-indicator type ball joint should be replaced. Technician B says that ball joints on both sides should be replaced even if one side only is worn. Who is correct?

   A.   A only

   B.   B only

   C.   Both A and B

   D.   Neither A nor B

   Answer A is correct. Only Technician A is correct. The ball joint's grease fitting serves as a wear indicator. The ball joint shows excessive wear and should be replaced.

   Answer B is incorrect. The ball joint on the opposite side of the vehicle does not have to be replaced, although it is likely to also be worn and should be carefully inspected.

   Answer C is incorrect. Only Technician A is correct.

   Answer D is incorrect. Only Technician B is incorrect.

30. Shock absorber mounts can be checked for good condition by using any of the following methods EXCEPT:

    A. inspecting them for dryness and cracking.

    B. observing them while bouncing the vehicle.

    C. performing a wheel alignment.

    D. watch the mounts while grasping the shock firmly and shaking it.

    TASK D.37

    Answer A is incorrect. A visual inspection is a valid way to check shock absorber mounts.

    Answer B is incorrect. Watching for looseness while bouncing the vehicle is an acceptable way to check the mounts.

    Answer C is correct. A wheel alignment would not necessarily reveal faulty shock absorber mounts.

    Answer D is incorrect. Grasping the shocks and shaking them is a good way to confirm that the shock mounts are OK.

31. The front caster setting on a McPherson strut equipped vehicle is found to be incorrect. Technician A says that there may be no OEM provision for making a caster adjustment. Technician B says that aftermarket parts may enable caster to be adjusted on a front-strut equipped vehicle. Who is correct?

    TASK D.42

    A. A only

    B. B only

    C. Both A and B

    D. Neither A nor B

    Answer A is incorrect. Technician B is also correct

    Answer B is incorrect. Technician A is also correct.

    Answer C is correct. Both Technicians are correct. OEM manufacturers generally do not make provisions for caster adjustments on front McPherson strut vehicles, but aftermarket kits are available for doing so.

    Answer D is incorrect. Neither Technician is incorrect.

TASK D.48

32. Technician A says that tire wear at the outside edge of a tire could be caused by too much negative camber. Technician B says that too much wear on both outer edges of a tire could be caused by over-inflation of the tire. Who is correct?

A. A only
B. B only
C. Both A and B
D. Neither A nor B

Answer A is incorrect. Negative camber would cause the inner edge of the tire to wear prematurely.

Answer B is incorrect. Over-inflation would wear the middle of the tire, not the edges.

Answer C is incorrect. Neither Technician is correct.

Answer D is correct. Neither Technician is correct.

TASK E.2

33. Technician A says that contaminated brake fluid could cause a spongy brake pedal. Technician B says that contaminated brake fluid could cause corrosion of the brake pads. Who is correct?

A. A only
B. B only
C. Both A and B
D. Neither A nor B

Answer A is correct. Only Technician A is correct. A spongy brake pedal could be caused by contamination or air in the brake fluid.

Answer B is incorrect. The brake pads are not exposed to the brake fluid.

Answer C is incorrect. Only Technician A is correct.

Answer D is incorrect. Only Technician B is incorrect.

TASK E.7

34. The most commonly used brake fluid is:

A. DOT 3.
B. DOT 4.
C. DOT 5.
D. silicone brake fluid.

Answer A is correct. DOT 3 is the most commonly used brake fluid.

Answer B is incorrect. DOT 4 is not as commonly used as DOT 3.

Answer C is incorrect. DOT 5 is not as commonly used as DOT 3.

Answer D is incorrect. Silicone brake fluid is actually incompatible with the other (Glycol) types mentioned and is not as widely used.

35. Technician A says that a brake drum may be machined until the dimension shown in the illustration above is reached. Technician B says that the drum should be inspected for hard spots and blue areas from excessive heat. Who is correct?

TASK E.9

     A. A only
     B. B only
     C. Both A and B
     D. Neither A nor B

     Answer A is incorrect. The dimension shown is the drum's maximum wear limit. If the drum diameter it is near to (or at) this dimension, it should be replaced, not machined. Some manufactures provide a "machine to" limit also.

     Answer B is correct. Only Technician B is correct. The drum should be inspected carefully for hard spots or blue spots, cracks and so forth. These would mean the drum should be scrapped.

     Answer C is incorrect. Only Technician B is correct.

     Answer D is incorrect. Only Technician A is incorrect.

36. The primary lubrication points of drum-type wheel brake assemblies include all of the following EXCEPT:

TASK E.11

     A. the raised pads on the backing plates.
     B. the interior surface of the brake drum.
     C. the contact points of self-adjusters.
     D. the star wheel brake adjuster.

     Answer A is incorrect. The pads should be lubricated.

     Answer B is correct. The interior of the brake drum should NOT be lubricated.

     Answer C is incorrect. All self-adjuster contact points should be lubricated.

     Answer D is incorrect. The star wheel brake shoe adjuster threads and shoe contact points should be lubricated.

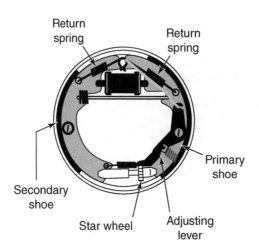

Return
spring

Return
spring

Primary
shoe

Secondary
shoe

Star wheel

Adjusting
lever

TASK E.14

37. Referring to the illustration above, at which location are the brakes initially adjusted by hand for proper brake operation?

A. The primary shoe

B. The secondary shoe

C. The star wheel

D. The adjusting lever

Answer A is incorrect. The primary shoe is not where the brakes are to be adjusted.

Answer B is incorrect. The secondary shoe is not where the brakes are to be adjusted.

Answer C is correct. The star wheel is where the brakes are to be initially adjusted.

Answer D is incorrect. The adjustment lever is part of the automatic brake adjusting mechanism.

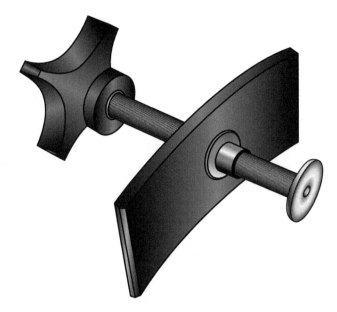

38. A disc brake caliper is being removed from a vehicle. Technician A says the tool shown above may be used to retract the caliper piston. Technician B says the tool shown above may be used to remove and install the brake pads. Who is correct?

TASK E.18

    A.   A only

    B.   B only

    C.   Both A and B

    D.   Neither A nor B

Answer A is correct. Only Technician A is correct. The tool shown is a caliper piston retracting tool

Answer B is incorrect. The tool is not used for removing or installing brake pads.

Answer C is incorrect. Only Technician A is correct.

Answer D is incorrect. Only Technician B is incorrect.

39. Brake rotor runout can best be checked by using a(an):

TASK E.22

    A.   inside micrometer.

    B.   dial indicator.

    C.   snap gauge.

    D.   feeler gauge.

Answer A is incorrect. An inside micrometer is not the correct measuring device.

Answer B is correct. Rotate the rotor while reading the dial indicator needle as it swings back and forth; compare to OEM specs.

Answer C is incorrect. A snap gauge is used for measuring the inside diameter of small holes.

Answer D is incorrect. A feeler gauge is used for measuring small gaps.

TASK E.25

40. Which of the following is LEAST LIKELY to help prevent brake pads from making a squealing noise?

    A.   Brake pad paste

    B.   Mechanic silencers

    C.   Bearing grease

    D.   Water- and heat-resistant lubricant

Answer A is incorrect. Brake pad paste may be applied to the back of brake pads to reduce vibration induced squeal.

Answer B is incorrect. Mechanical silencers help to stabilize brake pads and keep them from squealing.

Answer C is correct. Bearing grease cannot withstand the heat of disc brakes; it would likely contaminate them.

Answer D is incorrect. Some high-temperature dry lubricants are designed to help reduce vibration and squealing.

TASK E.27

41. Technician A says that on some vehicles pushing a button on the dashboard in the correct sequence will turn off the Brakes light. Technician B says that after fluid is added to a brake master cylinder, the "Brake Warning" light should go off. Who is correct?

    A.   A only

    B.   B only

    C.   Both A and B

    D.   Neither A nor B

Answer A is incorrect. Technician B is also correct

Answer B is incorrect. Technician A is also correct.

Answer C is correct. Both Technicians are correct. Some vehicles, for example, have a button by the dash used for resetting various "Service Soon" maintenance lights. Some vehicles have a "Low Brake Fluid" switch in the reservoir. Filling the master cylinder will open the switch and turn off the brake system warning light.

Answer D is incorrect. Neither Technician is incorrect.

TASK E.29

42. Burnishing new brake pads may require aggressively stopping a vehicle from 30 mph, at 30 second intervals, as many as:

    A.   10 times.

    B.   20 times.

    C.   30 times.

    D.   40 times.

Answer A is incorrect. This is not enough stops.

Answer B is incorrect. This is not enough stops.

Answer C is correct. This procedure is known in the trade as the 30/30/30 burnish procedure.

Answer D is incorrect. This is more stops than required.

43. The vacuum supplied to the power brake booster leaks out several minutes after the engine is shut off. Any of the following could be the cause EXCEPT for a leaking:

    A. brake booster.

    B. vacuum check valve.

    C. vacuum hose upstream of the check valve.

    D. vacuum hose downstream of the check valve.

    TASK E.31

    Answer A is incorrect. The booster could leak and could lose vacuum prematurely.

    Answer B is incorrect. The check valve could leak and cause a loss of vacuum.

    Answer C is correct. This hose is upstream of the check valve, so it would not be expected to hold a vacuum.

    Answer D is incorrect. This hose could leak vacuum since it is on the booster side of the vacuum check valve.

44. A vehicle experiences the MIL light turning on while driving. After connecting a scan tool, the technician finds several diagnostic trouble codes (DTCs) displayed at the same time. The most likely cause is:

    A. an incorrect voltage supply.

    B. an open circuit.

    C. a short circuit.

    D. a corroded ground connection.

    TASK F.2

    Answer A is incorrect. A supply voltage problem is not the most likely cause.

    Answer B is incorrect. An open circuit is not the most likely cause.

    Answer C is incorrect. A short would blow the circuit's fuse.

    Answer D is correct. When several DTCs suddenly show up on the scan tool at the same time, a common cause is a poor ground connection shared by the faulted circuits.

45. A fully charged 12-volt flooded cell automotive battery should have an open circuit voltage of

    A. 10.2 volts.

    B. 12.0 volts.

    C. 12.6 volts.

    D. 13.2 volts.

    TASK F.5

    Answer A is incorrect. 10.2 volts is too low to be fully charged.

    Answer B is incorrect. A so-called 12-volt battery should actually be called a 12.6 volt battery.

    Answer C is correct. Six 2.1 volt fully charged cells connected in series add up to 12.6 volts.

    Answer D is incorrect. 13.2 volts is a minimal charging system output.

TASK F.7

46. A flooded lead acid battery with removable caps should be only be topped off with:

    A. sulfuric acid.

    B. tap water.

    C. carbonated water.

    D. distilled water.

Answer A is incorrect. NEVER add more acid to a battery.

Answer B is incorrect. Tap water may contain impurities and additives (fluoride) and should not be added to a battery.

Answer C is incorrect. Do NOT add carbonated water to a flooded lead acid battery.

Answer D is correct. Use only distilled water to avoid contaminating the battery's electrolyte.

TASK F.12

47. A technician is not able to hear a manual transmission vehicle "click," or crank, when the ignition key is moved to the START position. Which of the following could be the most likely cause?

    A. A low battery

    B. Dirty battery connections

    C. A faulty neutral safety switch

    D. The clutch pedal not fully depressed

Answer A is incorrect. A low battery would at least cause a "clicking" sound of the cycling starter relay.

Answer B is incorrect. Dirty connections would not likely prevent even a "clicking" sound from being heard.

Answer C is incorrect. Neutral safety switches are used on vehicles with automatic transmissions.

Answer D is correct. Manual transmission vehicles commonly use a clutch pedal switch as a starting control circuit safety feature.

TASK F.14

48. To test the charging system in the field, it's OK to:

    A. remove the B+ battery cable from the battery while the engine is running to see if it stays running.

    B. rev the engine and see if a test light glows brighter.

    C. connect an ammeter across the battery and measure charging amp.

    D. check battery open circuit voltage when the engine is OFF, and compare it to when the engine is running at 2000 rpm.

Answer A is incorrect. This is not an approved procedure; it is risky, and it could ruin the alternator.

Answer B is incorrect. A test light is hardly a worthwhile test instrument.

Answer C is incorrect. NEVER connect an ammeter across a battery unless you want it ruined – or worse!

Answer D is correct. Taking open circuit voltage readings under various speed and load conditions is an acceptable procedure.

49. Fog lights fail to operate when the high beams are on but work OK with the low beams. Which of the following is the most likely cause?

TASK F.17

A. The high beam switch is defective.

B. The fog light switch is faulty.

C. The fog light fuse is blown.

D. Fog lights are not supposed to work with the high beams on.

Answer A is incorrect. It is unlikely that the switch is defective since the high beams work.

Answer B is incorrect. This is not the most likely cause because the fog lights work with the low beams.

Answer C is incorrect. Not so; the fog lights work with the low beams on.

Answer D is correct. It's against the law for fog lights to operate with the high beams on.

50. Technician A says a scan tool can be used to reset some vehicles' "Change Oil Soon" maintenance reminder lights. Technician B says that the menu for the on-board Driver Information Center used on some vehicles can be used to reset the oil change reminder. Who is correct?

TASK F.20

A. A only

B. B only

C. Both A and B

D. Neither A nor B

Answer A is incorrect. Technician B is also correct.

Answer B is incorrect. Technician A is also correct.

Answer C is correct. Both Technicians are correct. Both of these statements indicate correct ways to reset maintenance indicator lights.

Answer D is incorrect. Neither Technician is incorrect.

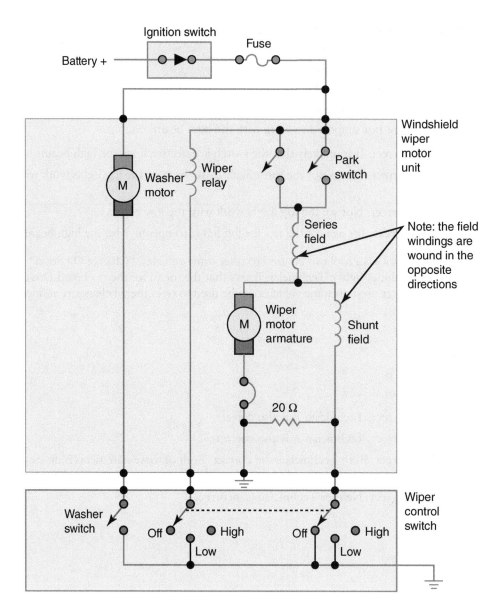

TASK F.22

51. Referring to the schematic diagram above, Technician A says that the washer motor can be tested by connecting a wire from the battery to the wiper relay input. Technician B says to ground the input terminal at the washer switch to make the washer motor operate. Who is correct?

A. A only

B. B only

C. Both A and B

D. Neither A nor B

Answer A is incorrect. Connecting power to the wiper motor would not test the washer motor.

Answer B is correct. Only Technician B is correct. Providing such a ground bypasses the possibly defective washer switch.

Answer C is incorrect. Only Technician B is correct.

Answer D is incorrect. Only Technician A is incorrect.

52. A leak detector is being used to check an A/C system. Technician A says to move the wand slowly along the top of all fittings, hoses, etc. Technician B says to use an open flame to check for leaks. Who is correct?

TASK G.2

   A. A only

   B. B only

   C. Both A and B

   D. Neither A nor B

Answer A is incorrect. The leak detecting wand should be moved along the bottom of A/C hoses and connections.

Answer B is incorrect. NEVER use an open flame to check for leaks. The reaction with refrigerant could generate a poisonous gas reaction.

Answer C is incorrect. Neither Technician is correct

Answer D is correct. Neither Technician is correct.

53. A cabin air filter is found to be clogged with debris. Technician A says that this could be the result of leaves getting into the cowling. Technician B says this could be caused by improper car washing. Who is correct?

TASK G.4

   A. A only

   B. B only

   C. Both A and B

   D. Neither A nor B

Answer A is correct. Only Technician A is correct. If the car is left outdoors, leaves could have gotten into the cowling and fallen into the air box.

Answer B is incorrect. Car washing might have gotten water into the cowling, but it would run out of the drain.

Answer C is incorrect. Only Technician A is correct

Answer D is incorrect. Only Technician B is incorrect.

54. An A/C "V" drive belt is being adjusted. The proper tension can be determined by using a:

TASK G.5

   A. ruler.

   B. scale.

   C. Vernier caliper.

   D. protractor.

Answer A is correct. A ruler can be used to measure belt deflection to then indicate if the belt is properly tensioned.

Answer B is incorrect. A scale would not be used for checking a "V" belt.

Answer C is incorrect. A Vernier caliper would not be used.

Answer D is incorrect. A protractor would not be useful for checking belt tension.

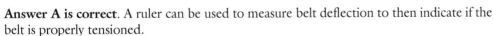

TASK G.6

55. A good way to clean an evaporator drain is to use:

A.   a stiff brush.

B.   a coat hanger.

C.   a shop vacuum.

D.   compressed shop air.

Answer A is incorrect. A stiff brush would be too large.

Answer B is correct. A bent-over piece of coat hanger wire makes a great tool for carefully clearing debris out of the condensate drain tube.

Answer C is incorrect. A Shop Vac could be tried, but would probably be too awkward.

Answer D is incorrect. Compressed air would blow the debris back into the air box.

Appendices

PREPARATION EXAM ANSWER SHEET FORMS

ANSWER SHEET

1. _____	20. _____	39. _____
2. _____	21. _____	40. _____
3. _____	22. _____	41. _____
4. _____	23. _____	42. _____
5. _____	24. _____	43. _____
6. _____	25. _____	44. _____
7. _____	26. _____	45. _____
8. _____	27. _____	46. _____
9. _____	28. _____	47. _____
10. _____	29. _____	48. _____
11. _____	30. _____	49. _____
12. _____	31. _____	50. _____
13. _____	32. _____	51. _____
14. _____	33. _____	52. _____
15. _____	34. _____	53. _____
16. _____	35. _____	54. _____
17. _____	36. _____	55. _____
18. _____	37. _____	
19. _____	38. _____	

ANSWER SHEET

1. _____	20. _____	39. _____
2. _____	21. _____	40. _____
3. _____	22. _____	41. _____
4. _____	23. _____	42. _____
5. _____	24. _____	43. _____
6. _____	25. _____	44. _____
7. _____	26. _____	45. _____
8. _____	27. _____	46. _____
9. _____	28. _____	47. _____
10. _____	29. _____	48. _____
11. _____	30. _____	49. _____
12. _____	31. _____	50. _____
13. _____	32. _____	51. _____
14. _____	33. _____	52. _____
15. _____	34. _____	53. _____
16. _____	35. _____	54. _____
17. _____	36. _____	55. _____
18. _____	37. _____	
19. _____	38. _____	

ANSWER SHEET

1. _____
2. _____
3. _____
4. _____
5. _____
6. _____
7. _____
8. _____
9. _____
10. _____
11. _____
12. _____
13. _____
14. _____
15. _____
16. _____
17. _____
18. _____
19. _____

20. _____
21. _____
22. _____
23. _____
24. _____
25. _____
26. _____
27. _____
28. _____
29. _____
30. _____
31. _____
32. _____
33. _____
34. _____
35. _____
36. _____
37. _____
38. _____

39. _____
40. _____
41. _____
42. _____
43. _____
44. _____
45. _____
46. _____
47. _____
48. _____
49. _____
50. _____
51. _____
52. _____
53. _____
54. _____
55. _____

ANSWER SHEET

1. _____	20. _____	39. _____
2. _____	21. _____	40. _____
3. _____	22. _____	41. _____
4. _____	23. _____	42. _____
5. _____	24. _____	43. _____
6. _____	25. _____	44. _____
7. _____	26. _____	45. _____
8. _____	27. _____	46. _____
9. _____	28. _____	47. _____
10. _____	29. _____	48. _____
11. _____	30. _____	49. _____
12. _____	31. _____	50. _____
13. _____	32. _____	51. _____
14. _____	33. _____	52. _____
15. _____	34. _____	53. _____
16. _____	35. _____	54. _____
17. _____	36. _____	55. _____
18. _____	37. _____	
19. _____	38. _____	

ANSWER SHEET

1. _____	20. _____	39. _____
2. _____	21. _____	40. _____
3. _____	22. _____	41. _____
4. _____	23. _____	42. _____
5. _____	24. _____	43. _____
6. _____	25. _____	44. _____
7. _____	26. _____	45. _____
8. _____	27. _____	46. _____
9. _____	28. _____	47. _____
10. _____	29. _____	48. _____
11. _____	30. _____	49. _____
12. _____	31. _____	50. _____
13. _____	32. _____	51. _____
14. _____	33. _____	52. _____
15. _____	34. _____	53. _____
16. _____	35. _____	54. _____
17. _____	36. _____	55. _____
18. _____	37. _____	
19. _____	38. _____	

ANSWER SHEET

1. _____	20. _____	39. _____
2. _____	21. _____	40. _____
3. _____	22. _____	41. _____
4. _____	23. _____	42. _____
5. _____	24. _____	43. _____
6. _____	25. _____	44. _____
7. _____	26. _____	45. _____
8. _____	27. _____	46. _____
9. _____	28. _____	47. _____
10. _____	29. _____	48. _____
11. _____	30. _____	49. _____
12. _____	31. _____	50. _____
13. _____	32. _____	51. _____
14. _____	33. _____	52. _____
15. _____	34. _____	53. _____
16. _____	35. _____	54. _____
17. _____	36. _____	55. _____
18. _____	37. _____	
19. _____	38. _____	

Glossary

Absorbed Glass Mat (AGM) battery See Valve Regulated Lead-Acid (VRLA) battery

Aiming pads Small projections on the front of some headlights to which headlight aligning equipment may be attached.

Air bag diagnostic monitor (ASDM) An automotive computer responsible for air bag system operation.

Airflow restriction indicator An indicator located in the air cleaner or intake duct to display air cleaner restriction by the color of the indicator window.

Alternating current flows in one direction and then in the opposite direction.

American Petroleum Institute (API) rating A universal engine oil rating that classifies oils according to the type of service for which the oil is intended.

Amperes A measurement for the amount of current flowing in an electric circuit.

Analog meter A meter with a pointer and a scale to indicate a specific reading.

Analog voltage signal A signal that is continuously variable within a specific range.

Angular bearing load A load applied at an angle somewhere between the horizontal and vertical positions.

Assembly Line Data Link (ALDL) The Assembly Line Data Link was used on OBD-I vehicles. The ALDL enabled a technician to connect to various on-board control units (modules) to read and retrieve stored data or perform tests. The ALDL is not uniform in pin arrangement nor location in the vehicle.

Atom The smallest particle of an element.

Automatic Transmission Rebuilders Association (ATRA) provides technical information to transmission shops and technicians.

Automotive dealership sells and services vehicles produced by one or more vehicle manufacturers.

Automotive Service Association (ASA) promotes professionalism and excellence in the automotive repair industry through education, representation, and member services.

Belleville spring A diaphragm spring made from thin sheet metal that is formed into a cone shape.

Bimetallic strip contains two different metals fused together that expand at different rates and cause the strip to bend.

Binary coding The assignment of numeric values to digital signals.

Blowby The amount of leakage between the piston rings and the cylinder walls.

Bolt diameter The measurement across the threaded area of the bolt.

Bolt length The distance from the bottom of the bolt head to the end of the bolt.

Bottom dead center (BDC) The piston position in an engine when the piston is at the very bottom of its stroke.

Brake fade Occurs when the brake pedal height gradually decreases during a prolonged brake application.

Brake pedal free-play The amount of brake pedal movement before the booster pushrod contacts the master cylinder piston.

Brake pressure modulator valve (BPMV) An assembly containing the solenoid valves connected to each wheel in an anti-lock brake system (ABS).

British thermal unit The amount of heat required to raise the temperature of 1 pound of water 1°F.

Caliper-actuated parking brake A disc brake system with an additional lever and corkscrew added to the existing caliper piston. When the emergency brake lever is applied, a corkscrew is forced against the caliper pistons and applies the brakes, independent of the hydraulic braking system.

Canceling angles are present when the vibration from one universal joint is canceled by an equal and opposite vibration from another universal joint.

Capacitive-conductive battery tester Rather than relying on load testing of 12 volt batteries, a non-invasive capacitive-conductive battery test method is used to inject an AC signal into the battery to measure the battery's internal resistance. Hand-held capacitive testers are commonly used to check the CCA of starter batteries.

Catalyst A material that accelerates a chemical reaction without being changed itself.

Catalytic converter A component mounted in the exhaust system ahead of the muffler to reduce hydrocarbon (HC), carbon monoxide (CO), and oxides of nitrogen (NOx) emissions.

Cell group A battery cell group contains alternately spaced positive and negative plates kept apart by porous separators.

Central port injection (CPI) A fuel injection system with a central port injector that supplies fuel to a mechanical poppet injector in each intake port.

Circuit breaker A mechanical device that opens and protects an electric circuit if excessive current flow is present.

Closed loop A computer operating mode that occurs when the engine is partially warmed up. In this mode, the computer uses the oxygen sensor signal to control the air-fuel ratio.

Closed loop flushing procedure flushes the refrigeration system with the system intact, without allowing refrigerant to escape to the atmosphere.

Clutch pedal free-play The amount of clutch pedal movement before the release bearing contacts the pressure plate release levers.

Cohesion The tendency of engine oil to remain on the friction surfaces of engine components.

Combination valve A brake system valve that contains a metering valve, proportioning valve, and a switch to operate the brake system warning light.

Compound A compound is a material with two or more types of atoms.

Compound gauge has the ability to read two values.

Compression ratio The relationship between combustion chamber volume with the piston at TDC and the volume with the piston at BDC.

Conicity A manufacturing defect in a tire caused by improperly wound plies that causes the tire to be slightly cone-shaped.

Connector position assurance (CPA) pin holds two wiring connectors together so they cannot be separated until the pin is removed.

Constant-running release bearing makes light contact with the pressure plate release levers even when the clutch pedal is fully upward.

Constant velocity (CV) joints transfer a uniform torque and a constant speed while operating at a wide variety of angles.

Conventional theory states that current flows from positive to negative through an electric circuit.

Coolant hydrometer A tester that measures coolant specific gravity to determine the antifreeze content of the coolant.

Corrosive A material that is corrosive causes another material to be gradually worn away by chemical action.

Coupling point occurs in a torque converter when the impeller pump, turbine, and stator begin to rotate together.

Critical speed The rotational speed at which a component begins to vibrate.

Cross counts The number of times the oxygen sensor signal switches from lean to rich in a given time period.

Crossflow radiator A radiator in which the coolant flows horizontally from one radiator tank to the opposite tank.

Cross threaded A defective thread condition caused by starting a fastener onto its threads when the fastener is tipped slightly to one side and the threads on the fastener are not properly aligned.

Current diagnostic trouble codes (DTCs) represent faults that are present during the diagnosis.

Data Link Connector (DLC) The Data Link Connector is a standardized 16 pin electrical connector used on OBD-II vehicles. The DLC enables a technician to connect to various on-board control units (modules) to read and retrieve stored data or perform tests. The DLC is typically located below the dashboard on the driver's side of the vehicle.

Detonation The sudden explosion of the air-fuel mixture in the combustion chamber rather than a smooth burning action.

Diagnostic procedure charts located in service manuals to provide the necessary diagnostic steps in the proper order to diagnose specific vehicle problems.

Diagnostic Trouble Code (DTC) A 'DTC' is a code logged in the vehicle's computer (PCM, TCM, EBCM, etc) indicating a fault has been found which would affect vehicle operation or emissions.

Digital meter A meter with a digital reading that indicates a specific value.

Digital voltage signal A voltage signal that is either high or low.

Diode trio A small device in some alternators that contains three diodes.

Direct current flows only in one direction.

Discard thickness dimension If the rotor is thinner than the minimum discard thickness dimension shown by law on the rotor, it has to be replaced. Laws prohibit a shop from turning a rotor past the manufacturer's brake rotor minimum thickness. So doing would cause the rotor to be too thin, leading to brake failure.

Disc-type return spring A beveled steel washer that flattens out when pressure is applied.

Distributor ignition (DI) An ignition system that uses a distributor to distribute the spark from the coil to the spark plugs.

Dog teeth A set of small gear teeth around the outer gear diameter with gaps between the teeth.

Downflow radiator A radiator in which the coolant flows vertically downward from one radiator tank to the opposite tank.

Dual overhead cam (DOHC) An engine with the intake and exhaust valves mounted in the cylinder heads and two camshafts are mounted on each cylinder head.

Electric/Electronic Power Steering (EPS) EPS uses an electric motor to assist the driver of a vehicle. Sensors detect the position and torque at the steering column, and a computer module applies assistive torque via the motor, which connects to either the steering gear or steering column. This allows varying amounts of assistance to be applied depending on driving conditions.

Electro Motive Force (EMF) the amount of electrical potential or pressure (voltage) found in a circuit.

Electrolyte A mixture of sulfuric acid and water in a lead acid battery.

Electromagnetic induction The process of inducing a voltage in a conductor by moving the conductor through the magnetic field or vice versa.

Electromagnetic pickup coil A pickup coil containing a permanent magnet surrounded by a coil of wire.

Electrons Negatively charged particles located on the various rings of an atom.

Electron theory states that electrons move from negative to positive through an electric circuit.

Electronic ignition (EI) An ignition system with a coil for each spark plug or pair of spark plugs.

Element An element is a liquid, solid, or gas with only one type of atom.

Engine coolant temperature (ECT) sensor A thermistor-type sensor that sends an analog voltage signal to the power-train control module (PCM) in relation to coolant temperature.

Enhanced evaporative system has the capability to check for leaks in the system and detect a 0.040-inch or 0.020-inch diameter leak depending on the vehicle model year.

Environmental Protection Agency (EPA) A United States government agency in charge of all aspects of environmental protection.

EVAP emission leak detector (EELD) produces a nontoxic smoke and blows this smoke into various components for leak-detection purposes.

Female-type quick disconnect coupling A coupling attached to the end of a shop air hose.An opening in the center of the coupling is inserted over the male part of the quick disconnect coupling. The female part of the coupling contains the locking and release mechanism that allows easy locking and release action with the male part of the coupling.

Fire and explosion data section Part of an MSDS that informs employees regarding the flash point of hazardous materials.

Fixed constant velocity (CV) joint does not allow any inward or outward drive axle movement to compensate for changes in axle length.

Flash diagnostic trouble codes (DTCs) are displayed by the flashes of the malfunction indicator light (MIL) in the instrument panel.

Flash programming The process of downloading computer software from a scan tool or PC into an onboard computer.

Flash-to-pass feature allows a driver to move the signal light lever forward enough to operate the headlights on high beam to indicate he or she is going to pass the vehicle in front.

Flex fan A cooling fan with flexible blades that straighten out as engine speed increases to reduce engine power required to turn the fan.

Flexible clutch allows some movement between the clutch facings and the hub.

Floating caliper A brake caliper that is designed so it slides sideways during a brake application.

Forward bias An electrical connection between a voltage source and a diode that results in current flow through the diode.

Four-channel antilock brake system (ABS) has a pair of solenoids for each wheel to modulate the brake system pressure individually at each wheel.

Fuel cells electrochemically combine oxygen from the air with hydrogen from a hydrocarbon fuel to produce electricity.

Full fielding The process of supplying full field current to an alternator to obtain maximum output from the alternator.

Fuse A component that protects an electric circuit from excessive current flow.

Gear ratio The ratio between the drive and driven gears.

Good conductor A good conductor has one, two, or three valence electrons that move easily from atom to atom.

Grade marks Radial lines on the head of a bolt in the USC system indicating the hardness of the steel in the bolt.

Graphing voltmeter A voltmeter that indicates voltage readings in graph form.

Grounded circuit An unwanted copper-to-metal connection in an electric circuit.

Halogen A term for a group of chemically treated nonmetallic elements including chlorine, fluorine, and iodine.

Helical gears have teeth that are cut at an angle to the center line of the gear.

High impedance test light A test light that contains a high resistance bulb.

History diagnostic trouble codes (DTCs) are caused by intermittent faults and these faults are not present during the diagnosis.

Hotchkiss drive An open drive shaft with two universal joints.

Hybrid organic additive technology (HOAT) A special coolant additive package to help prevent cooling system corrosion.

Hybrid vehicle A vehicle with two power sources such as a gasoline engine and an electric motor.

Hydrometer A tester that measures the specific gravity of a liquid.

Hygroscopic Having the ability to attract and hold water molecules. For example, DOT3 brake fluid is hygroscopic in that (too an extent), it will retain moisture which finds its way into the hydraulic brake system. Because of their affinity for adsorbing/absorbing moisture, hygroscopic materials should be stored in sealed containers.

Ignitable A substance that is ignitable can be ignited spontaneously or by another source of heat or flame.

Incandescence The process of changing electrical energy to heat energy in a light bulb to produce light.

Independent repair shop An independent repair shop is privately owned and operated without being affiliated with a vehicle manufacturer, automotive parts manufacturer, or chain organization.

Inductive pickup A type of pickup that senses the amount of current flow from the magnetic strength surrounding a conductor.

Infinite ohmmeter reading A resistance reading on an ohmmeter that is beyond measurement.

Injector pulse width The length of time in milliseconds that an injector is open.

In-phase universal joints are all on the same plane.

Insulator An insulator has five or more valence electrons which do not move easily from atom to atom.

International Automotive Technicians Network (iATN) A large group of automotive technicians in many countries that share technical knowledge with other members through the Internet.

International System A system of weights and measures in which every unit may be multiplied or divided by 10 to obtain larger or smaller units.

Jounce travel Upward tire and wheel movement.

Keep alive memory (KAM) The keep alive memory function stores data in the PCM, TCM or BCM when the vehicle's ignition is turned off. Data saved might relate to important engine operating parameters such as fuel trim, to mundane data such as radio station presets or power seat settings. When the battery is disconnected, typically KAM stored data is lost and must be reprogrammed.

Kinetic energy Energy in motion.

Laboratory (lab) scope A scope that displays various waveforms across the screen with a very fast trace.

Latent heat of condensation The heat released during condensation from a gas to a liquid.

Latent heat of vaporization The amount of heat required to change a liquid to a gas after the liquid reaches the boiling point.

Lateral tire runout The amount of sideways wobble in a rotating tire.

Left-hand thread A fastener with a left-hand thread must be rotated counterclockwise to tighten the fastener.

Line pressure The pressure delivered from the pressure regulator valve in an automatic transmission.

Male-type quick disconnect coupling The part of a coupling that is threaded into an air-operated tool. This male coupling contains grooves and ridges to provide a locking action with the female part of the coupling.

Malfunction indicator light (MIL) A warning light in the instrument panel that is illuminated by the PCM to indicate engine computer system defects.

Material Safety Data Sheets provide all the necessary data about hazardous materials.

Message center A digital display in the instrument panel where various warning messages are displayed.

Microprocessor The decision-making and calculating chip in a computer.

Mobile Air Conditioning Society Worldwide (MACS Worldwide) provides technical and business information to automotive air conditioning shops and technicians.

Molecule A molecule is the smallest particle that a compound can be divided into and still retain its characteristics.

Monolith-type catalytic converter contains a catalyst-coated monolith that is similar to a honeycomb.

Multiport fuel injection (MFI) A fuel injection system that opens two or more injectors simultaneously.

National Automotive Technicians' Education Foundation (NATEF) An independent affiliate of ASE responsible for evaluating and certifying automotive, autobody, and medium/heavy truck training programs.

National Institute for Automotive Service Excellence (ASE) An independent organization responsible for testing and certification of automotive, autobody, and medium/heavy truck technicians in the United States.

Neutrons Particles with no electrical charge that are located with the protons in the nucleus of an atom.

Normally closed contacts are closed when no pressure is supplied to the unit.

Normally open contacts are open when no pressure is supplied to the unit.

Occupational Safety and Health Administration (OSHA) regulates working conditions in the United States.

Ohms A measurement for resistance in an electric circuit.

On-board diagnostic II (OBD II) A type of computer system installed in all cars and light-duty trucks manufactured since 1996.

Open circuit An unwanted break in a electric circuit.

Open circuit voltage test A voltage test that is performed with no electrical load on the battery.

Open loop A computer operating mode that occurs during engine warmup. In this mode, the computer ignores the oxygen sensor signal and the computer program and other parameters control the air-fuel ratio.

Opposed cylinder engine An engine with the cylinder banks mounted at 180 degrees in relation to each other.

Out-of-round A variation in drum diameter at different locations.

Outside micrometer A precision measuring instrument designed to measure the outside diameter of various components.

Overhead cam (OHC) An engine with the intake valves, exhaust valves, and the camshaft mounted in the cylinder heads.

Overhead valve (OHV) An engine with the valves mounted in the cylinder heads and the camshaft and valve lifters located in the engine block.

Overrunning Alternator Decoupler (OAD) or Alternator Decoupler Pulley (ADP) Today's vehicles are factory-equipped with an Alternator Decoupler Pulley rather than a conventional solid pulley or a one-way clutch (OWC) pulley. The OAD/ADP is vehicle-specific, designed to produce quieter and smoother accessory drive operation and reduce vibrations in the passenger compartment. A decoupler pulley is similar in appearance to solid and One Way Clutch (OWC) pulley designs but plays a vital role in synchronizing the belt drive system. The OAD decouples the alternator from the rotational irregularities of the crankshaft and is only driven during the acceleration components of the rotating crankshaft.

Oxidation A process that occurs when some engine oil combines with oxygen in the air to form an undesirable compound.

Oxides of nitrogen (NOx) An exhaust emission caused by nitrogen and oxygen combining at high temperature above 2,500°F in the combustion chambers.

Pellet-type catalytic converter contains 100,000 to 200,000 small pellets coated with a catalyst material.

Periodic table A listing of the known elements according to their number of protons and electrons.

Permissible exposure limit (PEL) A section on MSDS sheets that indicates the maximum amount of hazardous material in the air to which a person may be exposed on a daily basis without harmful effects on the human body.

Personal protection equipment (PPE) Personal protective equipment, or PPE, is designed to protect workers from serious workplace injuries or illnesses resulting from contact with chemical, radiological, physical, electrical, mechanical, or other workplace hazards. Besides face shields, safety glasses, hard hats, and safety shoes, protective equipment includes a variety of devices and garments such as goggles, coveralls, gloves, vests, earplugs, and respirators.

Photochemical smog is formed by sunlight reacting with hydrocarbons (HC) and oxides of nitrogen (NOx).

Physical data section Part of an MSDS that provides information about the hazardous material, such as appearance and odor of the material.

Plus size or Plus sizing The act of outfitting/customizing a vehicle with larger than OEM (stock) tires.

Plunging constant velocity (CV) joint allows inward and outward drive axle movement to compensate for changes in axle length.

Ported manifold vacuum is supplied from above the edge of the throttle valve(s).

Positive crankcase ventilation (PCV) valve A valve that controls the amount of crankcase vapors flowing through the valve into the intake manifold.

Power booster performance test A series of mechanical and vacuum assisted tests to determine if the power booster is performing as it should.

Preignition The ignition of the air-fuel mixture in the combustion chamber by means other than the spark plug, such as a hot piece of carbon.

Printed circuit board A thin, insulating board used to mount and connect various electronic components such as resistors, diodes, switches, capacitors, and microchips in a pattern of conductive lines.

Protons Positively charged particles located in the nucleus of an atom.

Pulse width modulation (PWM) An on/off voltage signal with a variable on time.

Quad driver A group of four transistors in a computer chip that operates some of the output controls.

Quench area The area in the combustion chamber near the metal surfaces where the flame front is extinguished.

Quick-lube shop specializes in automotive lubrication work.

Radial bearing load A load applied in a vertical direction.

Radial runout The amount of diameter variation in a tire.

Reactive A material that is reactive reacts with some other chemicals and gives off a gas(es) during the reaction.

Reactivity and stability section Part of an MSDS that informs employees regarding the mixing of other materials with a hazardous material.

Rebound travel Downward tire and wheel movement.

Reciprocating An up-and-down or back-and-forth motion.

Resource Conservation and Recovery Act (RCRA) States that hazardous material users are responsible for hazardous materials from the time they become a waste until the proper waste disposal is completed.

Reverse bias An electrical connection between a voltage source and a diode that causes the diode to block voltage.

Right-hand thread A fastener with a right-hand thread must be rotated clockwise when tightening the fastener.

Rigid clutch Does not allow any movement between the clutch facings and the hub.

Rotary flow occurs when the fluid in a torque converter moves in a circular direction with the impeller pump and turbine.

Semiconductor A semiconductor has four valence electrons and these materials are used in the manufacture of diodes and transistors.

Separators are positioned between each pair of battery plates to keep these plates from touching each other.

Sequential fuel injection (SFI) An injection system in which each injector is opened individually.

Servo action occurs when the operation of the primary brake shoe applies mechanical force to help apply the secondary shoe.

Shorted circuit An unwanted copper-to-copper connection in an electric circuit.

Society of Automotive Engineers (SAE) rating A universal oil rating that classifies oil viscosity in relation to that atmospheric temperature in which the oil will be operating.

Solenoid An electro-mechanical device used to effect a push-pull mechanical operation using electric current.

Spark plug heat range indicates the ability of a spark plug to dissipate heat.

Spark plug reach The distance from the lower end of the plug shell to the shoulder on the shell.

Specialty Shop specializes in one type of repair work.

Specific gravity is the weight of a liquid, such as battery electrolyte, in relation to an equal volume of water.

Speed density system A fuel injection system in which the two main signals used for air-fuel ratio control are engine speed and manifold absolute pressure (MAP).

Sprung weight The weight of the chassis supported by the springs.

Stabilizer bar A long, spring steel bar connected from the crossmember to each lower control arm to reduce body sway.

Staked hub nut A nut that is secured to a drive axle by staking the nut lip into a recess in the axle.

Starter drive A mechanical device that connects and disconnects the starter armature shaft and the flywheel ring gear.

Static pressure The pressure in a system when the system is inoperative.

Stoichiometric air-fuel ratio The ideal air-fuel ratio of 14.7:1 on a gasoline engine that provides the best engine performance, economy, and emissions.

Straight-cut gears have teeth that are parallel to the gear centerline.

Strut chatter A clicking noise that is heard when the front wheels are turned to the right or left.

Supplementary Restraint System (SRS) Any of several vehicle occupant restraining systems, in addition to seatbelts, such as airbags.

Synchronizer brings two components to the same speed to allow shifting without gear clashing.

Synthetic lubricant is developed in a laboratory.

Technical service bulletins (TSB) Bulletins issued on a non-scheduled basis by original equipment vehicle manufacturers (OEMs) which alert service personnel of changes to repair procedures, parts updates and important information.

Thermistor A special resistor that increases its resistance as the temperature decreases.

Thermodynamics The relationship between heat energy and mechanical energy.

Thread depth The height of the thread from its base to the top of its peak.

Thread pitch In the USC system, thread pitch is the number of threads per inch.

Three-channel antilock brake system (ABS) has a pair of solenoids for each front wheel, but only one pair of solenoids for both rear wheels.

Threshold limit value A section on MSDS sheets that indicates the maximum amount of hazardous material in the air to which a person may be exposed on a daily basis without harmful effects on the human body.

Thrust bearing load A load applied in a horizontal direction.

Top dead center (TDC) The piston position in a engine when the piston is at the very top of its stroke.

Torque A twisting force.

Torque steer A condition in which the vehicle steering pulls to one side during hard acceleration.

Toxic A toxic substance is poisonous to animal or human life.

T-pin A common pin that is T-shaped.

Tracking bar is attached from the chassis to the rear suspension to prevent rear lateral body movement.

Transaxle vent is located in the transaxle case to allow air to escape from the transaxle.